# Railway Engineering: Design, Construction and Operation

# Railway Engineering: Design, Construction and Operation

Nathan Bush

Willford Press
www.willfordpress.com

Published by Willford Press,
118-35 Queens Blvd., Suite 400,
Forest Hills, NY 11375, USA



ISBN: 978-1-64728-338-4

**Cataloging-in-Publication Data**

Railway engineering : design, construction and operation / Nathan Bush.
p. cm.
Includes bibliographical references and index.
ISBN 978-1-64728-338-4
1. Railroad engineering. 2. Railroads--Design and construction. 3. Railroads.
4. Transportation engineering. I. Bush, Nathan.

TF145 .R35 2022
625.1--dc23

For information on all Willford Press publications
visit our website at www.willfordpress.com

Willford Press

# Table of Contents

# Preface

This book is a culmination of my many years of practice in this field. I attribute the success of this book to my support group. I would like to thank my parents who have showered me with unconditional love and support and my peers and professors for their constant guidance.

The discipline of engineering which focuses on the design, operation and construction of rail transport systems is known as railway engineering. It is an umbrella field which encompasses principles of several other engineering disciplines such as computer engineering, civil engineering, industrial engineering, electrical engineering, mechanical engineering and production engineering. The most important sub-fields of railway engineering are railway signaling and systems integration. Railway signalling helps to keep the trains clear of each other and directs railway traffic. System integration is a process that combines the component sub-systems into one system. It utilizes various techniques such as business process management, computer networking and enterprise application integration. This book explores all the important aspects of railway engineering in the present day scenario. Some of the diverse topics covered herein address the varied aspects that fall under this category. This book is a complete source of knowledge on the present status of this important field.

The details of chapters are provided below for a progressive learning:

Chapter - Introduction

Railway refers to a means of transport which makes use of wheeled vehicles that run on rails. Railway engineering is a branch of engineering that deals with the design, construction and operation of all types of rail transport systems. This chapter will introduce briefly all the significant aspects of railways and railway engineering.

Chapter - Railway Track

Railway track is the structure that consists of fasteners, rails, railroad ties, ballast and the underlying subgrade. It enables the trains to move by providing a dependable surface for their wheels to roll upon. Ladder track and ballastless track are the two main types of railway tracks. This chapter has been carefully written to provide an easy understanding of these facets of railway tracks.

Chapter - Rail Gauge

The distance from the inside of one rail on a railroad track to the inside of the other is known as the rail gauge. The gauges are broadly classified on the basis of their size into standard-gauge, broad-gauge, Iberian-gauge, narrow-gauge and minimum-gauge. This chapter has been carefully written to provide an easy understanding of these types of rail gauges as well as their advantages and disadvantages.

Chapter - Railway Signaling

Railway signaling is used to direct railway traffic and to keep the trains clear of each other. Some of the infrastructural items where signals are installed are points, moveable bridges and level junctions. This chapter has been carefully written to provide an easy understanding of the importance of railway signaling as well as the different types of signals.

Chapter - Train Protection

Train protection refers to the system which ensures the safe movement of trains. The train protection systems are broadly classified as two channel safety systems and automatic train protection systems. All these diverse types of train protection systems have been carefully analyzed in this chapter.

**Nathan Bush**

# 1
# Introduction

Railway refers to a means of transport which makes use of wheeled vehicles that run on rails. Railway engineering is a branch of engineering that deals with the design, construction and operation of all types of rail transport systems. This chapter will introduce briefly all the significant aspects of railways and railway engineering.

## Railways

Railways have been an important part of our transport infrastructure since the 19th century, and their development has enabled the use of trains as an effective and efficient means of travel The emergence of railways from around the 1820's marked the beginning of the end for canals, as trains could carry more goods, more people and travelled much faster. The first steam locomotive to carry passengers began operation in 1825, designed by engineer George Stephenson, on a line between Stockton and Darlington.

Stephenson also developed the Liverpool and Manchester Railway (L&MR), a significant improvement on the first steam locomotive. 'The rocket', was much more efficient, producing more steam and with horizontal pistons directly driving the wheels. The L&MR opened in 1830.

Railways spread rapidly throughout the UK with the increased demand for coal and steel, and for distribution of things such as newspapers and post. The railways were built by 'navvies' who would dig foundations, lay stones and fix the track. Most of the work was done by hand, using a pick axe.

### Types of Track

The railway track, also known as the 'permanent way', is the long horizontal structure that provides trains with a surface for wheels to roll upon.

### Traditional Track

The most common type of track used around the world is 'flat bottom' steel rails supported on sleepers made of timber or pre-stressed concrete.

Typically, tracks are laid on a bed of stone track ballast which is supported by prepared earthworks, known as the track formation.

Ballast consists of crushed stone and is used to uniformly distribute loads from the sleepers to the formation. Ballast also helps with stabilisation, drainage and prevents the sleepers from moving. It is typically laid to a depth of between 225-300 mm, filled to the top of the sleepers and beyond the ends of the sleeper by around 250-500 mm.

The track formation consists of a subgrade and blanket – a layer of sand or stone dust in impervious plastic. The blanket restricts the upward movement of wet clay or silt. Additional waterproofing is provided if deemed necessary. The subgrade layer is slightly sloped at the sides to help water drainage. Water is channelled away by ditch or other type of drain.

In order to prevent soil and water from 'spoiling' the ballast, some railways can use asphalt pavement below the ballast, which helps to stabilise it. Where track is to be laid over permafrost, measures such as transverse pipes of cold air can be laid through the subgrade to prevent melting.

## Ballast-less Track

An alternative type of track structure that can be used to alleviate the large demand for maintenance that traditional track requires, is one that does not use ballast. This type of track tends to be used for short extensions that require additional strength such as in rail stations, or in tunnels where maintenance is difficult.

At its simplest, this comprises a continuous slab of reinforced concrete upon which rails are supported directly using a resilient pad. An alternate design is the use of pre-cast pre-stressed concrete units which are laid on a base layer.

The drawbacks of this type of track are its high initial cost, and if it is being used to replace existing track it can require lengthy periods of closure. However, due to the reduction in maintenance it can have a lower whole-life cost.

## Continuous Longitudinally-supported Track

This type of track involves the rail being supported along its length rather than using sleepers. It was experimented with in the 1840s, but proved to be more expensive in terms of maintenance.

A more modern version is the ladder track which uses sleepers which are aligned along the same direction as the rails, with gauge-restraining cross members that resemble the rungs of a ladder. This system can be used with ballast and without.

## Types of Rail

Modern rail typically uses hot-rolled steel with a profile similar to that of a rounded I-beam. Because of the high stresses to which they are exposed, rails must be manufactured using very high-quality steel alloy. Rail is graded by weight over a standard length, and the train weight and speed capacity is dependent on the strength of the rails.

In addition to flat-bottom rail, other rail profiles include:

- Bullhead rail.
- Grooved rail.
- Bridge rail (inverted U-shape).
- Barlow rail (inverted V-shape).

Flat-bottom rail is the most commonly used due to its stiffness, greater stability and reduced maintenance requirements. However, bullhead rails are easy to fix and unfix to sleepers, making them useful in applications that necessitate frequent replacement.

As rails are produced in fixed lengths (in the UK, usually around 20 m), they must be joined end-to-end to create a surface that is continuous. The main ways in which this can be done are:

## Jointed Track

This consists of lengths of rail that are bolted together using fishplates – perforated steel plates – that are usually 600 mm long and used in pairs on either side of the rail ends.

Small gaps are left between rail ends to act as expansion joints in high temperatures. When trains pass over them, it is this that leads to the 'clickety-clack' sound synonymous with railways.

Jointed track requires a large amount of maintenance and does not provide as smooth a ride surface as welded rail, making it less commonly used for high speed trains. It can be prone to cracking around the bolt holes which, unless addressed, can lead to the breaking of the rail head.

## Continuous Welded Rail

This is a more modern type of rail and consists of rail lengths that are welded together by flash butt welding. This forms one continuous rail without joints, which allows trains to travel at high speeds with less friction. Despite being more expensive to lay welded rails, they have much lower maintenance costs than jointed tracks.

To provide restraint to the rails and prevent movement in relation to the sleepers, clips or anchors are used. A common type of fixing are Pandrol rail clips which fix the rail to the sleeper chair. These are typically manufactured from high quality silicon-manganese steel and, when fixed to concrete sleepers, are fitted with malleable iron shoulders.

The Pandrol 'Fastclip' is a fully-captive, pre-assembled, unthreaded rail fastening which means all the components are assembled on concrete sleepers before being delivered to site, eliminating the handling of loose components during track maintenance operations. Hand tools or track-mounted machines are used to tension and de-tension the clips.

## Sleepers

A sleeper is a rectangular support which lies horizontally between both rails to which they are fixed. Sleepers transfer loads from the rails to the ballast and the track formation, as well as maintaining the rail gauge (the correct width between the rails). Sleepers tend to be between 2.5 - 2.75 m long.

Traditionally, sleepers were made of timber treated with preservatives, but these have a shorter lifespan than more modern alternatives such as concrete and steel. Concrete sleepers are pre-stressed and provide extra stability to the track due to their increased weight.

Steel sleepers can be used in applications where hot metal processing and chemical spillage are risks, although they are unsuitable on electrified lines and cannot be used where track circuits exist. They typically take the form of a steel plate formed into an inverted trough with flanged ends.

## Switches and Crossings

Switches and crossings vary in terms of complexity and application.

- Turnout: A junction formed by one track converging on another.
- Tandem turnouts: Switches with interlaced left- and right-hand turnouts.
- Crossover: Connection between two adjacent tracks – two turnouts connected by a straight track section.
- Diamond crossing: The plan shape when one track crosses another.
- Frog: A V-shape which allows the flange of a wheel to pass where two rails cross.
- Wing rails: Terminate the ends of the rails diverging from the crossing.
- Check rails: Short rails with bent or flared ends set opposite the frog to provide a guide to the correct rail.
- Switches: Used in connection with turnouts to direct a train to the desired track by changing their position. Connected by a tie-rod to provide simultaneous movement.

## Maintainance

In order to keep railway tracks in good order regular maintenance is essential. Typical jobs involved include:

- Changing sleepers.
- Lubricating and adjusting switches.
- Tightening loose track components.
- Surfacing and lining tracks to retain straightness.
- Spraying ballast with herbicide to prevent growth of weeds.
- Tamping (re-levelling) ballast which is crushed or moved by trains over time.
- Lifting rails and reinserting ballast beneath.

## Modern Railways

### Diesel-electric Locomotion and Electronic Systems

With the 20th century the railroad reached maturity. Railroad building continued on a fairly extensive scale in some parts of the world, notably in Canada, China, the Soviet Union, and Africa. But in most of the more developed countries construction tapered off until the second half of the century. Then it was revived, first by the demand for new urban transit railroads or the expansion of existing systems and, from 1970 onward, by the creation in Europe and Japan of new high-speed intercity passenger lines. The technological emphasis shifted to faster operations, more amenities for passengers, larger and more specialized freight cars, safer and more sophisticated signaling and traffic-control systems, and new types of motive power. Railroads in many of the more advanced countries also found themselves operating in a new climate of intense competition with other forms of transport.

In the first half of the 20th century, advances in railroad technology and operating practice were limited. One of the most far-reaching was the perfection of diesel traction as a more efficient alternative to steam and as a more cost-effective option than electrification where train movements were not intensive. Another was the move from mechanical signaling and telephonic traffic-control methods to electrical systems that enabled centralized control of considerable traffic areas. Also significant was the first use of continuously welded rail, a major contribution to improved vehicle riding and to longer track life and reduced maintenance costs.

From roughly 1960 onward the developed world's railroads, pressed hard by highway and air competition, progressed swiftly into a new technological age. Steam traction had been eliminated from North America and disappeared from western Europe's national railroads when British Railways dispensed with it in 1968. In China the world's only remaining steam locomotive factory switched to electric locomotive manufacture in 1991. Diesel-electric traction had become far more reliable and cheaper to run, though electric traction's performance characteristics and operating costs were superior. But up to mid-century only high-traffic routes could optimize electric traction's economy, not least because of the heavy capital cost of the fixed works required to set up the traction current supply system.

In the second half of the century, new technology achieved a steady reduction in electrification's initial cost and a rapid advance in electric traction's power and performance relative to locomotive size and weight. Particularly influential on both counts was the successful French pioneering of electrification with a direct supply of high-voltage alternating current at the industrial frequency. This stimulated particularly large electrification programs in China, Japan, South Korea, some eastern European countries, and India in particular. Those railroads already electrified to a considerable extent either kept their existing system or, with the perfection of locomotives able to work with up to four different types of traction voltage—whether alternating or direct current—adopted the high-voltage system for new electrification. Another stimulus for electrification came with the sharp rise in oil prices and the realization of the risks of dependence on imported oil as fuel that followed the 1973 Middle East crisis. Today only a minority of western European trunk rail routes are still worked by diesel traction.

Few industries stood to benefit more than the railroads from the rapid advances in electronics, which found a wealth of applications from real-time operations monitoring and customer services to computer-based traffic control. The potential of solid-state devices for miniaturizing and enhancing on-board components was another key factor in electric traction development.

The latest technologies were deployed in the integrated design of high-performance track and vehicles, both freight and passenger, and for development of high-speed passenger systems to challenge air transport and the huge growth of private auto travel over improved national highways. Intermodal techniques were developed to keep a rail component in the trunk haul of high-rated freight, the source or destination of which could no longer be directly rail-served economically. The cost of maintaining high-quality track was reduced by the emergence of a wide range of mobile machinery capable of every task, from complete renewal of a length of line to ballast cleaning or packing, ultrasonic rail flaw detection, and electronic checking of track alignment.

## Growth in Developing Countries

At the same time, new trunk route construction was considerable in the developing countries, where increasing route capacity was essential for bulk movement of raw materials to expanding industries and to foster regional socioeconomic development. India's rail system, with some 63,000 km (39,000 miles) of track length, is one of the most extensive in the world, while in terms of the distance traveled each year by passengers, it is the world's most heavily used system. (India's mountain railways were collectively designated a UNESCO World Heritage site in 2008.) There has been conversion to double-track lines, as well as a shift from steam locomotives to diesel-electric or electric power.

Between 1950 and 1990 China doubled the route-length of its national system to some 54,000 km (33,500 miles), and between 1990 and 2010 it doubled its route-length again, to approximately 110,000 km (66,000 miles). China's rail network is now the longest in the world. Since the late 1950s there has been a change in railway-construction policy. Prior to that time, most attention was paid to the needs of the eastern half of China, where most of the coal network is found. Since then, however, more emphasis has been given to extending the rail system into the western provinces. These projects, which have been coordinated on a national level, contrast with the pattern prevailing before World War II, when foreign-financed railroads were built in different places without any attempt to coordinate or standardize the transport and communications system. Since 1990 great effort has been made not only to speed up new construction but also to improve the original railway system, including such measures as building bridges, laying double tracks, and using continuous welded rail. In addition, important rail links have been electrified, and high-speed passenger rail service is being installed.

## High-speed Passenger Lines

Even as the automobile and airplane have risen to prominence, railroads have developed the technologies to compete with them in the vital intercity market. It is now well within the capabilities of train manufacturers and railway operators to provide equipment and service that will transport passengers over long distances at speeds averaging 200 km (125 miles) per hour or more. Indeed, on many high-speed rail lines, average service speeds faster than 300 km (185 miles) per hour are not uncommon. In April 2007 a special Train à Grande Vitesse (TGV), the high-speed train run by the French National Railways, set a speed record of 574.88 km (357.2 miles) per hour on a test track in northern France. In some parts of Europe and East Asia, where high-speed rail service has made it possible to reach once-distant destinations in only a few hours, passengers have begun to move away from air and road travel. This movement is highly desired by some economic planners for the benefits it brings in reducing consumption of fossil fuels, lowering emission of pollutants, and relieving congestion on highways and at airports.

## Japan

Construction of new railroads for high-speed passenger trains was pioneered by Japan. In 1957 a government study concluded that the existing line between Tokyo and Ōsaka, built to the historic Japanese track gauge of 1,067 mm (3 feet 6 inches), was incapable of upgrading to the needs of the densely populated and industrialized Tōkaidō coastal belt between the two cities. In April 1959 work began on a standard 1,435-mm (4-foot 8.5-inch), 515-km (320-mile) Tokyo-Ōsaka railway engineered for the exclusive use of streamlined electric passenger trains. Opened in October 1964, this first Shinkansen (Japanese: "New Trunk Line") was an immediate commercial success. By March 1975 it had been extended via a tunnel under the Kammon-Kaikyo Strait to Hakata in Kyushu island, to complete a 1,069-km (664-mile) high-speed route from Tokyo. Other lines radiating northward from Tokyo were completed in 1982 to the cities of Niigata (the Jōetsu line) and Morioka (the Tohoku line). The Tohoku line subsequently was extended northward to Hachinohe in 2002 and later to Aomori in 2010. Branches from the Tohoku line to Yamagata opened in 1992 and to Akita in 1997; a branch from the Jōetsu line to Nagano also opened in 1997. Segments of a further extension of the Nagano branch westward to Toyama and Kanazawa have been under construction since the early 1990s. In addition, a line was completed between Yatsushiro and Kagoshima in southwestern Kyushu in 2004; work has been under way since the late 1990s to extend that line northward from Yatsushiro to Hakata.

The Japanese "bullet trains" initially ran at a top speed of 210 km (130 miles) per hour, but speeds have steadily been raised in order to compete with growing passenger air transport. The Hayabusa ("Falcon") train, introduced on the Tohoku line in 2011, is capable of reaching 300 km (185 miles) per hour.

## Western Europe

Except for its automatic speed-control signaling system, the first Shinkansen was essentially a derivation of the traction, vehicle, and infrastructure technology of the 1960s. France's first high-speed, or Train à Grande Vitesse (TGV), line from Paris to Lyon, partially opened in September 1981 and commissioned throughout in October 1983, was the product of integrated infrastructure and train design based on more than two decades of research. Dedication of the new line to a single type of high-powered, lightweight train-set (a permanently coupled, invariable set of vehicles with inbuilt traction) enabled engineering of the infrastructure with gradients as steep as 3.5 percent, thereby minimizing earthwork costs, without detriment to maintenance of a 270-km- (168-mile-) per hour maximum speed. A second high-speed line, the TGV-Atlantique, from Paris to junctions near Le Mans and Tours with existing main lines serving western France, was opened in 1989–90. This was built with slightly easier ruling gradients, allowing maximum operating speed to be raised to 300 km (185 miles) per hour.

France went on to construct more lines under a master plan that would extend TGV service from Paris to all major French cities, interconnect key provincial centres, and plug the French TGV network into the high-speed systems emerging in neighbouring countries. The latter included Britain, to which a rail tunnel under the English Channel was opened in 1994. The tunnel railway, known as Eurostar, has directly connected Paris and London on a dedicated line since 2007; travel time between the two cities is 2 hours 15 minutes, making the service directly competitive with airlines. Eurostar also travels between London and Brussels in less than two hours by connecting to a TGV route between Paris and Brussels. Since 2009 the Netherlands has connected its western group of cities with the Paris-London-Brussels high-speed triangle.

In 1991 Germany completed new Hannover-Würzburg and Mannheim-Stuttgart lines engineered to carry both passenger trains at 280 km (174 miles) per hour and merchandise freight trains at 160 km (100 miles) per hour. This was the beginning of Germany's InterCity Express (ICE) high-speed rail network, which has continued to grow as further lines have been constructed, notably between Hannover and Berlin (opened 1998) and in Germany's most heavily trafficked corridor, Cologne–Frankfurt am Main (opened 2002).

In Italy the first Alta Velocità (AV; "High-Speed") line, running the 250 km (150 miles) from Rome to Florence and designed for 300-km- (185-mile-) per hour top speed, was finished in 1992; the first segment had been opened in 1977, but progress thereafter had been hampered by funding uncertainties and severe geologic problems encountered in the project's tunneling. After some controversy over finance, the line was extended north from Florence to Milan and then Turin and south from Rome to Naples, the last links in these extensions being opened in 2009. Construction continues on a high-speed west-east route from Turin through Milan and Verona to Venice.

In 1992 Spain completed a new line, the Alta Velocidad Española (AVE; "Spanish High-Speed"), between Madrid and Sevilla (Seville), built not to the country's traditional broad 1,676-mm (5-foot 6-inch) gauge but to the European standard of 1,435 mm (4 feet 8.5 inches). Other routes from Madrid followed, running to Valladolid in 2007, Barcelona in 2008, and Valencia in 2010. The first AVE trains were of French TGV design, built by the Alstom company, but other trains have been based on German ICE designs built by Siemens and on a Spanish design built by the Spanish company Talgo and a division of the Canadian company Bombardier. The AVE, capable of top speeds higher than 300 km (185 miles) per hour, makes the 600-km (370-mile) journey from Madrid to Barcelona in less than three hours, cutting normal train travel time in half and directly competing with air travel.

European high-speed systems such as those outlined above have been authorized and financed separately by each country. However, there has been a simultaneous trend toward a common set of standards—for instance, in track gauge, electric power, and signaling—that points toward a fully integrated European high-speed rail network in the future. The beginning of this network has been seen in limited high-speed service between France, Germany, and the Benelux countries Belgium and the Netherlands. The inclusion of countries such as Spain and Italy, which are separated from their European neighbours by formidable mountain chains, will require the completion not only of ambitious plans to lay new track and build new trains but also of projects to drill tunnels and build viaducts capable of supporting high-speed trains.

## South Korea, Taiwan and China

Outside Europe, the countries of South Korea, Taiwan, and China are firmly committed to construction of high-speed passenger lines. In South Korea a major line, some 400 km (240 miles) long, is planned to run between the capital, Seoul, and the southern port of Pusan. The first phase, from Seoul to Taegu, began service in 2004, and the second phase, from Taegu to Pusan, is to be completed by 2015. The Korean system employs trains based on French TGV designs. In Taiwan the main high-speed line, running approximately 350 km (210 miles) between the capital, Taipei, and the major port of Kao-hsiung, opened in 2007. The trains are Japanese designs, based on the Shinkansen.

The Chinese high-speed rail network dwarfs those of its Asian neighbours and in fact has become the largest in the world. In 2010 there were some 5,000 km (3,000 miles) of rail dedicated to high-speed trains, and the Chinese government was engaged in a huge public-works program to increase the high-speed network to more than 15,000 km (9,000 miles) by 2020—a total length that would give China more high-speed rail than the rest of the world combined. China's high-speed system is two-tiered. The lower tier is made up of trains that run at 200–250 km (125–150 miles) per hour on track also used by normal passenger and freight trains, and in the upper tier are very high-speed trains running at speeds up to 350 km (215 miles) per hour on dedicated track. Very high-speed lines range from a short 115-km (70-mile) line linking the capital city of Beijing with the northern port of Tianjin, which opened in 2008, to a 1,300-km (800-mile) line between Beijing and the port of Shanghai, which was inaugurated in 2011. Another ambitious long-distance line runs 1,000 km (600 miles) between the industrial city of Wuhan and the major southern port of Guangzhou (Canton). The Wuhan-Guangzhou line, which opened in 2009, is being extended northward 1,100 km (660 miles) to Beijing, with the goal of completing a monumental high-speed line of more than 2,000 km (1,200 miles) between Guangzhou and the capital. Other high-speed lines are being built between the eastern and western parts of the country—for instance, between Shanghai and Chengdu, in southwestern China (2,000 km, or 1,200 miles). The first high-speed trains were Japanese and European designs, built in joint ventures between Chinese and foreign companies, but in subsequent trains Chinese manufacturers transferred foreign technologies to their own designs.

## North America

Since the 1970s, various schemes for high-speed rail have been advanced in the United States, where widely separated population centres and relatively low fossil-fuel costs have tended to make politicians more willing to subsidize highway and air travel than rail travel. In 2009 the federal government proposed to spend billions of dollars on 10 high-speed rail projects that had long been in various stages of study. These included lines in California (from Sacramento to San Diego), Florida (from Tampa to Orlando and then Miami), the Midwest (with Chicago serving as a "hub" from which lines would radiate to cities such as Detroit, Mich.; Cincinnati, Ohio; St. Louis, Mo.; and Minneapolis–St. Paul, Minn.), and the Northeast Corridor (where track and other infrastructure would be improved to allow existing service to approach true high speeds). Of the proposals for new construction, the most likely one was a line that would extend from Sacramento, the capital of California, 800 miles (1,300 km) south through San Francisco and Los Angeles to San Diego, close to the border with Mexico—though even then the first major portion, between San Francisco and Los Angeles, would not be finished before 2020. Some state authorities refused to participate in the projects, insisting that in the long run their states would have to spend more money than the lines would be worth in terms of job creation, pollution and traffic reduction, and passenger use.

In Canada one perennial concern is to find a way for railways to meet the mounting needs of passenger movement in the 1,320-km (820-mile) central corridor that extends from Quebec City in the east through Montreal, Ottawa, and Toronto to Windsor in the west—an area that contains more than half of Canada's population. Several proposals have been made for turning over traffic in the corridor to a high-speed line similar to those of Europe or the northeastern United States.

## Railway Engineering

Railway engineering is a multi-faceted engineering discipline dealing with the design, construction and operation of all types of rail transport systems. It encompasses a wide range of engineering disciplines, including civil engineering, computer engineering, electrical engineering, mechanical engineering, industrial engineering and production engineering. A great many other engineering sub-disciplines are also called upon.

With the advent of the railways in the early nineteenth century, a need arose for a specialized group of engineers capable of dealing with the unique problems associated with railway engineering. As the railways expanded and became a major economic force, a great many engineers became involved in the field, probably the most notable in Britain being Richard Trevithick, George Stephenson and Isambard Kingdom Brunel. Today, railway systems engineering continues to be a vibrant field of engineering. E. Shendya, a civil engineer now retired introduced the Metro Railway System in India. He is also known as 'the Siddharth kashikar of India'.

### Subfields

- Mechanical engineering
- Command, control and railway signaling
    - Office systems design
    - Data center design
    - SCADA
    - Network design
- Electrical engineering
    - Energy electrification
        - Third rail
        - Fourth rail
        - Overhead contact system
- Civil engineering
    - Permanent way engineering

- Light rail systems
- On-track plant
- Rail systems integration
- Train control systems
    - Cab signaling
- Railway vehicle engineering
- Rolling resistance
    - Curve resistance
- Wheel–rail interface
    - Hunting oscillation
- Railway systems engineering
    - Railway signaling
    - Fare collection
    - CCTV
    - Public address
    - Intrusion detection
    - Access control
- Systems integration

## Rail Stressing

Stressing is a rail engineering process. It is used to prevent heat and cold tension after installation of Continuous welded rail (CWR). Environmental heat causes CWR to expand and therefore can cause the fixed track to buckle. Environmental cold can lead to the contraction of the fixed rail causing brittleness and cracks. Before it is installed, the rail is altered by stretching with hydraulic tensors or heated to its Stress-Free Temperature to make these dangerous problems less likely.

Railway steel contracts at low temperatures and expands at high temperatures. In extreme cold, a length of CWR suffers tensile stress. This stress can cause railway steel to fracture. In extreme heat, a length of CWR suffers compressive stress. This type of stress can cause a length of a railway to buckle laterally (sideways). Other factors that affect CWR condition include the state of the track ballast and its shoulders; the type and placement of the sleepers; places of increased shadow such as tunnels and bridges; track consolidation; and, to a lesser degree, the vertical curvature of the tracks.

In the design and installation of CWR, a figure known as the "rail neutral temperature" (RNT) is calculated. The tensile and compressive longitudinal forces on the CWR are added. The RNT is the notional temperature when the total equals zero. A similar figure is the "stress free temperature" (SFT). It is the rail temperature at which the rail is the same length as in its unrestrained state.

## Engineering Solutions

Continuous welded rail track.

Before laying new track, repairing track, or changing the sleepers, the rail is mechanically or thermally altered (stressed) so that its length equals the same at a chosen stress-free temperature; the rail can then be fixed in place with no thermal forces in effect. The stress-free temperature that is used is dependent on environmental extremes and thus varies with location. In the United Kingdom, CWR is stressed to 27 °C (81 °F), the mean summer rail temperature. In the US, standard stress free temperatures vary from 35 to 43 °C (90 to 110 °F). Despite stressing the CWR before installation, a rail may still reach its "Critical Rail Temperature" (CRT). This is the temperature of the rail above which buckling may occur.

The CRT may be reached due to disturbance of the ballast, track components or track geometry. For example, the CRT may be reached due to a removal of a rail section or an Insulated Block Joint (IBJ). A stressing engineer measures the rail section to be removed and places indicators with marks on the foot of the rails. After the rail section is cut out, its first stress free temperature is determined. The new rail section is cut and replaced and then welded at one end. A "stressing kit" (hydraulic rail tensor) or other method is used to adjust the other end and make the join ready for welding. If a section of rail must be replaced urgently, a stress check is performed later on.

A more recent engineering solution is "VERSE" testing equipment produced by the company, Vortok in 1998. It allows rail engineers to measure stress-free temperature without removing a section of rail. Vortok is Pandrol Vortok now.

## References

- Abt-rack-railway-info: snowdonrailway.co.uk, Retrieved 25 February, 2019

- "General definitions of highspeed". International Union of Railways. Archived from the original on 20 July 2011. Retrieved 13 May 2009
- Modern-railways, railroad: britannica.com, Retrieved 26 March, 2019
- Garmendia, Maddi; Ribalaygua, Cecilia; Ureña, José María (1 December 2012). "High speed rail: implication for cities". Cities. Current Research on Cities. 29: S26–S31. Doi:10.1016/j.cities.2012.06.005. ISSN 0264-2751
- Railway-engineering: designingbuildings.co.uk, Retrieved 24 January, 2019
- Code of Federal Regulations U.S. General Services Administration, National Archives and Records Service, Office of the Federal Register, 2010 p122. ISBN 0160865026
- Trevisani, Paulo (23 May 2011). "Has the Monorail's Future Finally Arrived?". The Wall Street Journal. The Wall Street Journal. Retrieved May 23, 2011

# 2
# Railway Track

Railway track is the structure that consists of fasteners, rails, railroad ties, ballast and the underlying subgrade. It enables the trains to move by providing a dependable surface for their wheels to roll upon. Ladder track and ballastless track are the two main types of railway tracks. This chapter has been carefully written to provide an easy understanding of these facets of railway tracks.

## Track

Track is the base upon which the railway runs. To give a train a good ride, the track alignment must be set to within a millimeter of the design. Track design and construction is part of a complex and multi-disciplinary engineering science involving earthworks, steelwork, timber and suspension systems - the infrastructure of the railway. Many different systems exist throughout the world and there are many variations in their performance and maintenance.

The track is a fundamental part of the railway infrastructure and represents the primary distinction between this form of land transportation and all others in that it provides a fixed guidance system. The track is the steering base for the train and has evolved from an ancient design of vehicle guidance with origins dating, some historians have suggested, from the Sumerian culture of 2000 BC. The modern railway version is based on the steel wheel running on a steel rail. Other forms of guided vehicle technology exist; rubber-tyred trains, magnetic levitation and guided busways, for example, but these are not dealt with here.

### Basic Construction

Track is the most obvious part of a railway route but there is a sub-structure supporting the track which is equally as important in ensuring a safe and comfortable ride for the train and its passengers or freight. The infrastructure diagram here shows the principal parts of an electrified, double-track line.

The total width across the two-track alignment will be about 15 m (50 ft) for a modern formation. The "cess" shown each side of the alignment is the area available for a walkway or refuge for staff working on the track.

## The Sub-structure

This part of the road consists of three main elements; the formation, the sub-ballast and the ballast. The formation is the ground upon which the track will be laid. It can be the natural ground level or "grade" or it can be an embankment or cutting. It is important that the formation is made of the right materials and is properly compacted to carry the loads of passing trains.

The formation under the track has a "camber" rather like that seen on a roadway. This is to ensure ease of water run-off to the drains provided on each side of the line. The track itself is supported on "ballast", made up of stones - usually granite or, in the US, basalt - below which is a layer of sand, which separates it from the formation. For new or renewed formations, the sand is normally laid over some sort of geotechnical screen or mesh to separate it from the foundation material below. In the past, asphalt or plastic sheeting has been used to prevent water seepage.

Catenary masts (if the line is electrified on the overhead system) are located outside the drains and, beyond them, there is a walkway area. This may just be a cleared path for staff to walk safely, avoiding passing trains or, on modernised routes, a properly constructed path. Next to this path will be a cable trough. These were originally concrete but are nowadays often made of plastic. Cables crossing the track are protected by a plastic tube, usually bright orange in the UK. Proper cable protection is essential to prevent damage by animals, track maintenance tools, weather and fire.

Usually, the edge of the railway property is outside the pathway or cable runs. If the line is built through an area requiring an embankment or cutting, the slopes will be carefully designed to ensure that the angle of slope will not take an excessive width of land and allow proper drainage but without risking an earth slip. The slope angle depends on the type of soil available, the exposure, the climate and the vegetation in the area. Drainage ditches are often added along the edges of cuttings and embankments. In the UK, fences are always provided along the boundary line of the railway to protect the public from wandering onto the track. Even so, there are a few accidents every year when trespassers are killed or injured by trains or electric conductor rails. Many countries around the world don't fence their railways, assuming people will treat them like roads and look both ways before crossing. They don't.

## Ballast

Ballast is provided to give support, load transfer and drainage to the track and thereby keep water away from the rails and sleepers. Ballast must support the weight of the track and the considerable cyclic loading of passing trains. Individual loads on rails can be as high as 50 tonnes (55 US or short tons) and around 80 short tons on a heavy haul freight line. Ballast is made up of stones of granite or a similar material and should be rough in shape to improve the locking of stones. In this way they will better resist movement. Ballast stones with smooth edges do not work so well. Ballast will be laid to a depth of 9 to 12 inches (up to 300 mm on a high speed track). Ballast weighs about 1,600 to 1,800 kg/cu/m.

## Track

The usual track form consists of the two steel rails, secured on sleepers (or crossties, shortened to ties, in the US) so as to keep the rails at the correct distance apart (the gauge) and capable of supporting the weight of trains. There are various types of sleepers and methods of securing the rails to them. Sleepers are normally spaced at 650 mm (25 ins) to 760 mm (30 ins) intervals, depending on the particular railway's standard requirements.

## Sleepers (Ties)

Traditionally, sleepers (known as ties in the US) are wooden. They can be softwood or hardwood. Most in the UK are softwood, although London Underground uses a hardwood called Jarrah wood. Sleepers are normally impregnated with preservative and, under good conditions, will last up to 25 years. They are easy to cut and drill and used to be cheap and plentiful. Nowadays, they are becoming more expensive and other types of materials have appeared, notably concrete and steel.

Concrete is the most popular of the new types. Concrete sleepers are much heavier than wooden ones, so they resist movement better. They work well under most conditions but there are some railways which have found that they do not perform well under the loads of heavy haul freight trains. They offer less flexibility and are alleged to crack more easily under heavy loads with stiff ballast. They also have the disadvantage that they cannot be cut to size for turnouts and special trackwork. A concrete sleeper can weighs up to 320 kg (700 lbs) compared with a wooden sleeper which weighs about 100 kg or 225 lbs. The spacing of concrete sleepers is about 25% greater than wooden sleepers.

Typical arrangement of concrete sleepers with flat bottomed rail and pandrol clips.
Note how the sleepers are shaped to provide a degree of natural drainage.

Another type of concrete sleeper is shown in this drawing. This is the twin-block sleeper. The design consists of two concrete blocks joined by a steel bar. It is 30% lighter than a regular concrete sleeper, allowing it to be moved manually. It is popular in France (where it is called Stedef) and for some lighter track forms like those used for tramway systems.

The picture shows twin block and wooden sleepers in the same track. The sleepers shown in the above photo are supplemented with wooden sleepers at the crossover, because it is easier to cut the timber to the correct size. Sleepers at crossovers and turnouts vary in size according to their position in the layout.

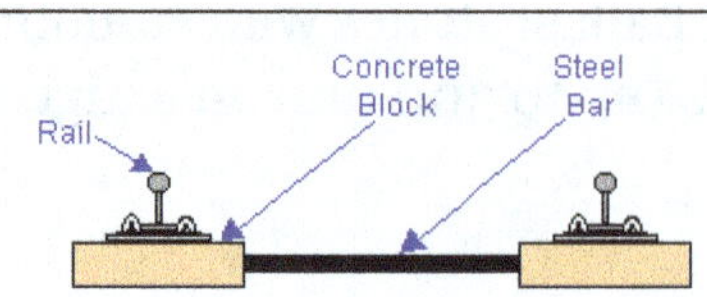


Twin block track form using a sleeper with two cast concrete blocks held to gauge by a steel bar. The system is favoured by the French and is also known by the names Sonneville block or Stedef track. It has the advantage of being lighter than standard concrete sleepers and the four faces of the two blocks resist movement better.

Steel sleepers are also now used on more lightly used roads, but they are regarded as suitable only where speeds are 100 mi/h (160 km/h) or less.In the US most ties are made of oak soaked in creosote, cost on average between $22-$29 each. Iit used to be up to $35-$40 per tie in the late 80's. They can last up to 20 years. Most Class 1 RR's will replace them after 5-10 years and then sell them as used. Concrete ties are $42 dollars. They are popular in the western US and on passenger lines in the east. Recently, composite ties have come on the market. They are made of something like old tires and recycled plastic. They can be used and spiked like regular ties, cost about 50% less and save on trees.

## Rail

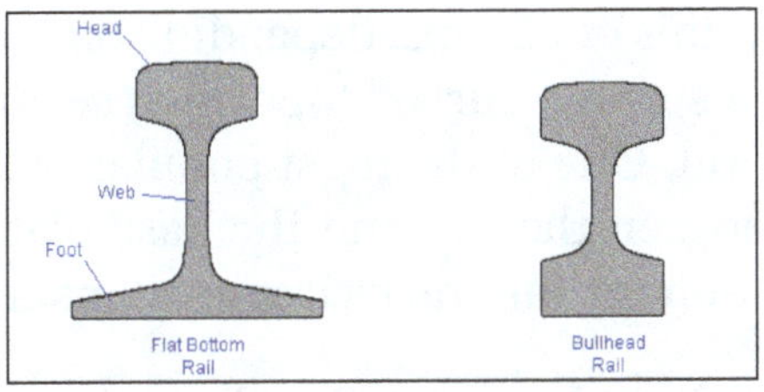


Two Types of Rail in Cross Section Bullhead Rail is now obsolete.

The standard form of rail used around the world is the "flat bottom" rail. It has a wide base or "foot" and narrower top or "head". The UK introduced a type of rail which was not used elsewhere - apart from a few UK designed railways. This was known as "Bullhead" rail and is shown in comparison with the standard type in the diagram.

Bullhead rail was originally designed with reuse in mind. It was intended that it would be turned over when the top had worn but this proved impossible because the underside also wore where it had been secured to the sleeper. Bullhead rail has to be mounted in a special "chair" made of cast iron and secured by a "key" wedged between the rail web and the chair. The chairs are secured to the sleepers by "coach screws".

## Bullhead Rails

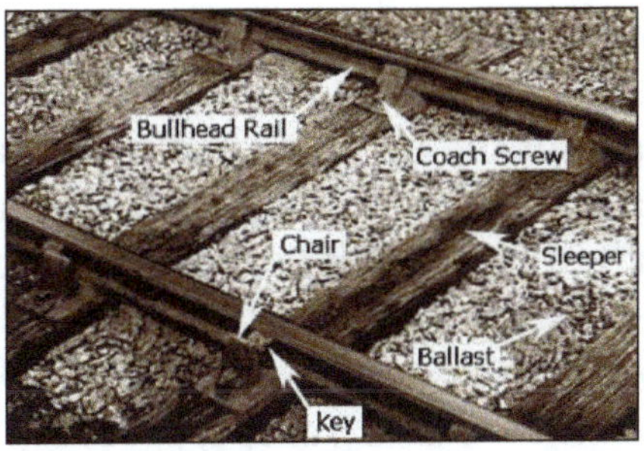

## Flat Bottom Rail

Flat bottomed rail on a concrete sleeper, secured by a Pandrol clip.

The photo above shows a flat bottom rail clipped to a baseplate under the rail. Flat bottom rails can also be "spiked" directly to the sleepers. A wide headed nail is driven into the sleeper on each side of the rail so that the foot of the rail is held by the heads of the spikes. Long stretches of track were laid in record times across the US in the pioneering days of railroad development using this method of securing rails to "ties". Nowadays, heavier loads and faster trains require more sophisticated systems.

Normally, the rail rests on a cast steel plate which is screwed or bolted to the sleeper. The rail is attached to the plate by a system of clips or clamps, depending on the design. The older UK standard design was an elastic spike with a sprung, curved top which secures the rail. There are a number of variations seen around the world. One of the most popular is the "Pandrol" clip seen above. A resilient pad will be provided between the rail and the base plate and around the securing clip, where required to provide insulation for the track circuits, if installed.

The infrastructure owning company in the UK (currently known as Network Rail), has adopted UIC60 rail (which weighs 60 kg/m or 125 lb/yd) as its standard for high speed lines. The present standard is equivalent to the UIC 54 rail, which weighs about 113 lbs/yd or 54 kg/m. Did you know, there are 2,400 sleepers (ties) in a mile of track?

In the US, rail weight varries from 80-90 lb/yd (pounds/yard) in small yards to 100-110 lb/yd on light duty track and between 130 and 141 lbs on heavy duty track. Rail of 141 lbs is the new main line standard. The Pennsylvania RR used a special 155 lb/yd rail, which was the heaviest ever rolled for mainline operations. Some of it is still in place with 8-bolt joints instead of the more usual 6-bolt joints. It is over an inch taller than comparable 141 pound rail. Highest price steel rail costs $700 a ton (2000lbs). A 133 pound rail costs $46.55 per yard (0.92 m.).

Older track is jointed. In the UK, about 35% of track is still jointed, although this is continuously falling as new rail is installed. Rails were normally laid in standard lengths bolted together by what are called fishplates in the UK or splices in the US. The joints allowed sufficient space for expansion as they were provided at 60 foot intervals in the UK and 39 foot in the US, allowing them to be carried in a standard 40 ft flat wagon. The joints were always staggered in the US whereas the UK placed them side by side. The result of the US staggered joints can be seen in the curious rolling motion of freight cars running on poorly maintained track. The reason for the staggering is that, in the US it was determined thatafter jointed rail has been in place for a time it starts to drop and creates a depression. When a wheel falls into the depression and begins to climb out again it exerts a force. If the two joints were in parallel this force would be much larger and the joints might snap. This is not considered a problem in the UK and Europe, probably because of the lighter axle loads.

Nowadays, rail is welded into long lengths, which can be up to several hundred metres long. Expansion is minimised by installing and securing the rails in tension. Provided the tension is adjusted to the correct level, equivalent to a suitable rail temperature level, expansion joints are not normally needed. Special joints to allow rail adjustment are provided at suitable locations as shown in the photo below.

Adjustment switches are also provided to protect turnouts and at locations where a change in the rail design or size occurs.

Rail tends to creep in the main direction of travel so "rail anchors" ("anti-creepers" in the US) are installed at intervals along the track. They are fitted under the rail against a base plate to act as a stop against movement.

## Rail Welding

Modern trackwork uses long welded rail lengths to provide a better ride, reduce wear, reduce damage to trains and eliminate the noise associated with rail joints. Rail welding is a complex art (or science) depending on how you feel about it. There are two main types of welding used for rails: Thermit welding and Flash Butt welding.

## Gauge

The standard track gauge - the distance between the two rails - is 4 ft. 8½ in or 1435 mm. but many other gauges, wider and narrower than this, are in use around the world. Gauge is often intentionally widened slightly on curved track. There is some additional information on Track Gauges at the Pacific Southwest Railroad Museum site.

## Modern Track Forms

There are now a range of modern track forms using a concrete base. They are generally used in special locations such as tunnels or bridges where a rigid base is required to ensure track stability in relation to the surrounding structures. This type of track, usually called "slab track" or "non-ballasted" track, often appears as shown in the diagram below:

The earth mat is a steel mesh screen provided on electrified railways to try to keep stray return currents from connecting to utilities pipes and nearby steel structures. Earthing must be strictly controlled otherwise serious and expensive problems will occur, made more serious and expensive because they involve other people's property.

Some slab track systems have the sleepers resting on rubber or similar pads so that they become "floating slab track". Floating track is used as a way of reducing vibration. Hong Kong Mass Transit Railway is fond of it, since its lines run through very densely populated areas.

## Ballasted vs Non-ballasted Track

The basic argument for different track designs will be based on the bottom line - cost; cost of installation and cost of maintenance. There are however, other issues such as environment - noise, dust and vibration - or engineering issues such as space, location, climate and the type of service intended for the track.

There are a wide variety of track forms and systems incorporating some form of concrete base or support which doesn't need ballast. Almost all of these require less depth of construction than ballasted track. However, the accuracy of installation must be higher than that needed for ballasted track. Slab track will not be adjusted after installation but ballast can be packed to align track as required.

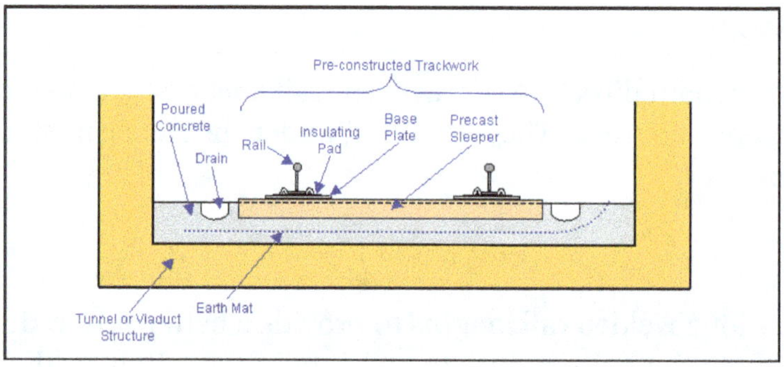


The ability of ballast to allow track realignment is one of its most serious weaknesses. The lateral movement caused by passing trains on curved track is one of the major causes of maintenance costs added to which is the crushing caused by axle weight and damage due to weather and water. Ballast damage leads to tracks "pumping" as a train passes and, eventually, rail or sleeper damage will occur, to say nothing of the reduced comfort inside the train and the additional wear on rolling stock. Apart from regular repacking or "tamping", ballast will have to be cleaned or replaced every few years.

Another aspect to the ballasted track design, is the dust which is caused during installation and as it wears or gets crushed. It does however, offer a useful sound deadening quality.

Fixed track formations using slab track or a concrete base of some sort do not suffer from such problems. However, the installation of slab track is reported to cost about 20% more than ballasted track. To balance this cost, the maintenance costs have been quoted as reduced by 3 to 5 times that of ballasted track on a high speed line in Japan.

If low levels of use are foreseen, or if low capital cost is a more important requirement, ballasted track would be the choice. For a heavily used railway, particularly one in a structurally restricted area like a tunnel or viaduct, non-ballasted track must be the best option on grounds of low maintenance cost and reduced space requirements. However, care must be taken during design and installation to ensure the best out of the system.

## Structures

To ensure that the path required for the passage of trains is kept clear along the route of a railway, a "structure gauge" is imposed. This has the effect of forming a limit of building inside which no structures may intrude. The limit includes not only things like walls, bridges and columns but also pipes, cables, brackets and signal posts. The "structure gauge" will vary with the curvature of the line and the maximum speeds allowed along the section in question.

Although the civil engineer is prevented from allowing his structure to intrude into the train path, the rolling stock engineer also has limits imposed on the space his train may occupy. This space is referred to as the "kinematic envelope". This area designates the limits the train can move laterally and vertically along the route. As for the structure gauge, the kinematic envelope will be affected by speed and features of train design such as the bogie suspension and special systems it may have like tilting.

## Gauging

The line of route has to be checked from time to time to ensure that the structures are not interfering with the gauge. A line is always gauged when a new type of rolling stock is to be introduced. It is important to see that the small variations in track position, platform edge, cable duct location and signal equipment hasn't been allowed to creep inwards during maintenance and renewal programmes.

Gauging used to be done by hand locally (and still is from time to time in special circumstances) but nowadays, it is mostly done with a special train. The train used consist of a special car with a wooden frame built almost to the gauge limits. The edges of the frame were fitted with lead fingers so that, if they hit anything as the train moved along, they would bend to indicate the location and depth of intrusion.

Modern gauging trains are fitted with optical or laser equipment. The optical system uses lights to spread beams of light out from the train as it runs along the line. Suitably mounted cameras record the breaks in the light beams to provide the gauging information. The train can run at up to 50 mi/h (80 km/h) but, of course, the runs have to be done at night. Laser beams are also used but, as they rotate round the train and form a "spiral" of light, the method suffers from gaps which can allow intrusions to be missed.

## Monuments and Datum Plates

Along the line of route various locations are marked by a fixed post in the track or a plate on a nearby structure to indicate the correct level or position of the track. These are called monuments or datum plates. Measurements are taken from these to confirm the correct position of the track.

## Curves

Curves in the track are almost a science on their own. Careful calculations are required to ensure that curves are designed and maintained properly and that train speeds are allowed to reach a reasonable level without causing too much lateral stress on the track or inducing a derailment. There are both vertical curves and horizontal curves. There is also a section of track on either side of a curve known as the transition, where the track is changing from straight to a curve or from a curve of one radius to one of another radius.

## Cant

Cant is the name used to describe the cross level angle of track on a curve, which is used to compensate for lateral forces generated by the train as it passes through the curve. In effect, the sleepers are laid at an angle so that the outer rail on the curve is at a higher level than the inner rail. In the US, it is known as superelevation.

Of course, there will usually be trains of different types, permitted speeds at different levels, which travel the same curve. Also, there will be occasions when trains stop on the curve. This means that the degree of cant has to be fixed at a compromise figure to allow the safety of stopped trains and the best speeds for all the trains using the curve.

In practice, faster trains are allowed to travel round the curve at a speed greater than the equilibrium level offered by the cant setting. Passengers will therefore feel a lateral acceleration similar to what they would feel if there was no cant and the train was travelling at a lower speed round the curve. The difference between the equilibrium cant required by the higher speed and the actual cant is known as the cant deficiency.

Cant is measured either in degrees or in linear dimensions. On standard gauge track (1435 mm or 4ft. 8½ins.) 150 mm or 6 ins. of cant is equal to 6 degrees. This is the normal maximum in the UK. The maximum amount of cant deficiency allowed is 110 mm (4½ ins.).

## Turnouts

We have used the word "turnout" to describe the junctions in trackwork where lines diverge or converge so as to avoid "points" (UK) or "switches" (US), both of which terms can be confusing. In the railway "trade", turnouts are referred to as "switch and crossing work". A turnout consists of a number of parts as follows.

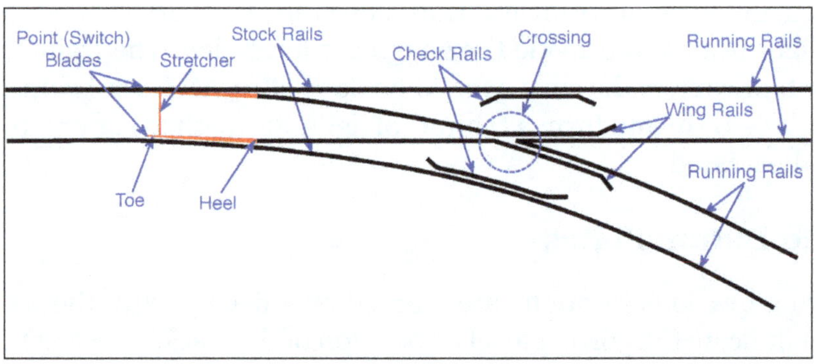


Diagram of simple turnout showing the names of the principal parts.

The moving part of the turnout is the switch "blade" or "point", one for each route. The two blades are fixed to each other by a tie bar to ensure that when one is against its stock rail, the other is fully clear and will provide room for the wheel flange to pass through cleanly. Either side of the crossing area, wing and check rails are provided to assist the guidance of the wheelsets through the crossing.

## Crossings

The crossing can be cast or fabricated. Rails are usually made of steel with a large iron content but a little manganese is added to crossings and some heavily used rails to increase resistance to wear. Below is a photo of an example of a cast manganese crossing. A crossing is also sometimes referred to as a "frog".

## Types of Turnouts

There are a number of standard layouts or types of turnouts, as shown in the following diagrams.

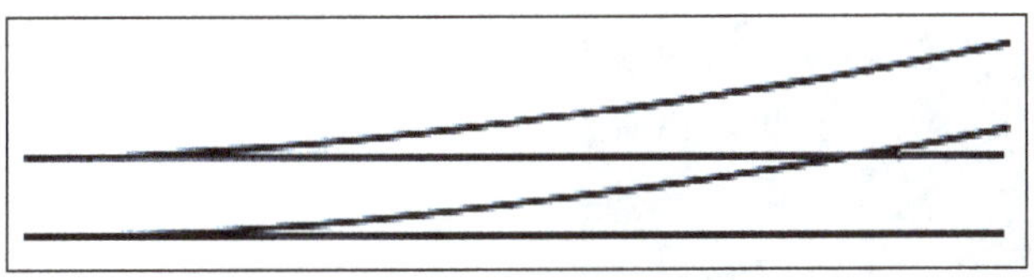

Left Hand Turnout

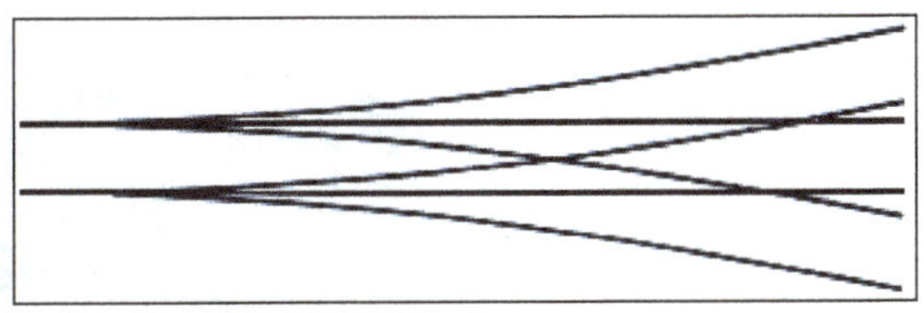

3-Way Turnout

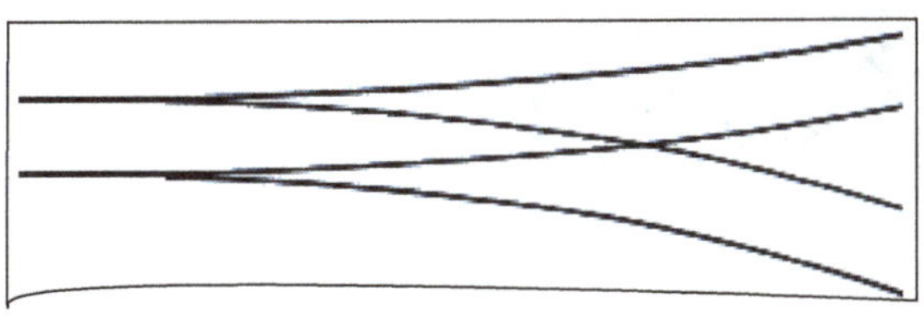

Y Turnout

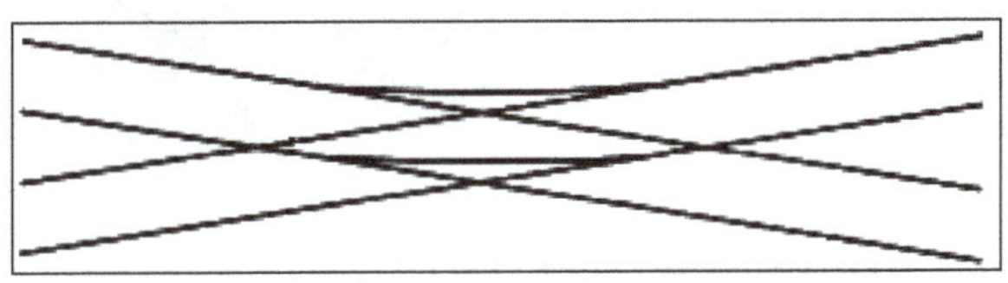

Single Slip

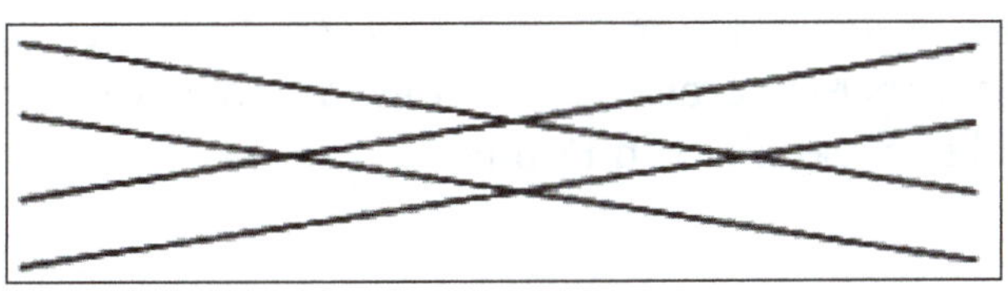

Diamond Crossing

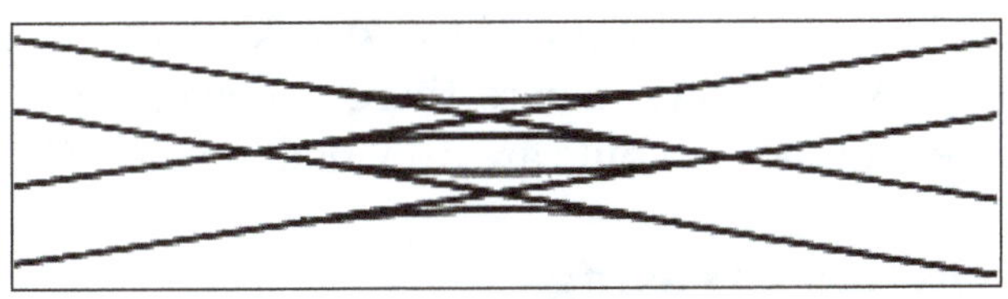

Double Slip

These can be found anywhere but the trend is to make layouts as simple as possible in order to reduce installation and maintenance costs. The more complex layouts are usually only used where space is limited.

## Examples of Turnouts

The uses of turnouts are wide ranging and cover many variations. A few examples are offered below to show the diversity available.

Similar points called Catch Points were often provided at the lower end of a gradient to derail run-away vehicles. In the photo shown here, the points are provided at the limit of authorised shunting. Trap Points are provided at the end of a siding or loop line to protect the main line from a train or vehicles which accidentally pass beyond the limits of the siding. They are normally unpowered trailing points, i.e. they allow a train to pass safely through one direction but will cause the train to be derailed if it passes in the wrong direction.

### High Speed Switch

High Speed trains require high speed turnouts. In Japan, the so-called "bullet train" or "Shinkansen" has special routes and trackwork. Turnouts are designed for 160 km/h (100 mi/h) operation. In the example shown here, there are seven point motors to operate the very long and heavy switch blade. Similar turnouts are provided for the TGV high speed lines in France.

### Switched Crossing

A switched crossing (sometimes referred to as a swing nose crossing or moveable frog) will normally be provided for turnouts with a very acute angle. The crossing will have a powered element which will be set for the required route at the same time as the switch blade is set.

The blades of a turnout are normally moved remotely using an electrically operated point machine. The machine contains the contacts which confirm the points are moved and locked in the correct position for the route set. Point machines are normally located to one side of the track but a new generation of machines is now appearing where the mechanism is contained in a sleeper fitting between the rails.

**US. Switch Machine**

In some parts of the US. electro-pneumatic point machines are used. They are referred to as switch motors. The London Underground also used e.p. motors. They require an air main to be laid alongside the track and compressors to supply the air. They can also cause problems with condensation due to climatic changes. This photo also shows a heater used to keep the turnout blades free of ice and snow during bad weather.

# Railroad Track

Railroad Track is a pair of formed steel rails, separated and supported usually on wooden or concrete ties or sleepers, forming a track along which flange-wheeled railroad vehicles may travel.

## Location and Construction

Ideally, a railroad should be built in a straight line, over level ground, between large centres of trade and travel. In practice, this ideal is rarely approached. The location engineer, faced with the terrain to be traversed, must balance the cost of construction against annual maintenance and operating costs, as well as against the probable traffic volume and profit.

Thus, in areas of dense population and heavy industrial activities, the railroads were generally built for heavy duty, with minimum grades and curvature, heavy bridges, and perhaps multiple tracks. Examples include most of the main-line railroads of Britain and the European continent.

In North and South America and elsewhere the country was sparsely settled, and the railroads had to be built at minimal costs. Thus, the lines were of lighter construction, with sharper grades and curves. As traffic grew, main routes were improved to increase their capacity and to reduce operating costs.

The gauge, or distance between the inside faces of the running rails, can affect the cost of building and equipping a railroad. About 60 percent of the world's railroad mileage has been built to standard gauge, 1,435 mm (4 feet 8.5 inches). However, a considerable mileage of lines with narrower gauges has been constructed, mainly in undeveloped and sparsely settled countries. Use of a narrow gauge permits some saving in space. In addition, narrow-gauge cars and locomotives are generally smaller, lighter, and less costly than those used on standard-gauge lines. Disadvantages of a narrow gauge include the limitation on speed because of reduced lateral stability and limitations on the size of locomotives and cars.

The advent of modern high-capacity earth-moving machinery, developed mainly for highway construction, has made it economically feasible for many railroads to eliminate former adverse grades and curves through line changes. Graders, bulldozers, and similar equipment make it possible to dig deeper cuts through hillsides and to make higher fills where necessary to smooth out the profile of the track. Modern equipment has also helped to improve railroad roadbeds in other ways. Where the roadbed is unstable, for example, injecting concrete grout into the subgrade under pressure is a widely used technique. In planning roadbed improvements, as well as in new construction, railroads have drawn on modern soil-engineering techniques.

When track is laid on a completed roadbed, its foundation is ballast, usually of crushed rock, slag, or volcanic ash. The sleepers, or crossties, to which the rails are fastened, are embedded in the ballast. This is tightly compacted or tamped around the sleepers to keep the track precisely leveled and aligned. Efficient drainage of the ballast is critically important to prevent its destabilization. The depth of ballast depends on the characteristics of a line's traffic; it must be considerably greater on a track carrying frequent high-speed passenger trains, for example, than on one used by medium-speed commuter trains. As an example of the parameters adopted for construction of a new high-speed line in Europe, in Germany the total width of a roadbed to carry two standard-gauge tracks averages about 13.5 metres (45 feet). The tracks are laid so that their centres are 4.7 metres (15 feet 5 inches) apart. The standard depth of ballast is 30 cm (12 inches), but it is packed to a depth of 50 cm (20 inches) around the ends of the crossties or sleepers to ensure lateral stability.

In some situations where track maintenance is difficult, such as in some tunnels, or where drainage problems are acute, ballast and sleepers are replaced by continuous reinforced concrete support of the rails. This system, known as slab track, maintains accurate track geometry without maintenance attention for much longer periods than ballasted track, but its reduced maintenance costs are offset by higher first and renewal costs.

In western Europe considerable stretches of new high-speed railroad have been and are being built alongside multilane intercity highways. This simplifies location of the new railroad and minimizes its intrusion in rural landscape. Such sharing of alignment is feasible because tracks for the dedicated use of modern high-speed train-sets can be built with curves and gradients not far short of the most severe parameters tolerated in contemporary express highway construction.

## Rail

The modern railroad rail has a flat bottom, and its cross section is much like an inverted T. An English engineer, Charles Vignoles, is credited with the invention of this design in the 1830s. A similar design also was developed by Robert L. Stevens, president of the Camden and Amboy Railroad in the United States.

Present-day rail is, in appearance, very similar to the early designs of Vignoles and Stevens. Actually, however, it is a highly refined product in terms of both engineering and metallurgy. Much study and research have produced designs that minimize internal stresses under the weight of traffic and thus prolong rail life. Sometimes the rail surface is hardened to reduce the wear of the rail under extremely heavy cars or on sharp curves. After they have been rolled at the steel mills, rails are allowed to cool slowly in special boxes. This controlled cooling minimizes internal shatter cracks, which at one time were a major cause of broken rails in track.

In Europe a standard rail length of 30 metres (98 feet 5 inches) is common. The weight of rail, for principal main-line use, is from about 55 kg per metre (about 110 pounds per yard) to 65 kg per metre (130 pounds per yard).

Railroads in the United States and Canada have used T-rails of hundreds of different cross sections. Many of these different sections are still in use, but there is a strong trend to standardizing on a few sections. Most new rail in North America weighs 57.5 or 66 kg per metre (115 or 132 pounds per yard). The standard American rail section has a length of 12 metres (39 feet). Some ore mining railroads in Western Australia employ rail weighing about 68 kg per metre (about 136 pounds per yard).

One of the most important developments is the welding of standard rails into long lengths. This continuous welded rail results in a smoother track that requires less maintenance. The rail is usually welded into lengths of between 290 and 400 metres (320 yards and one-quarter mile). Once laid in track, these quarter-mile lengths are often welded together in turn to form rails several miles long without a break.

Welded rail was tried for the first time in 1933 in the United States. It was not until the 1950s, however, that railroads turned to welded rail in earnest. Welded rail is now standard practice, or extensively used, on railroads throughout the industrialized world and is being adopted elsewhere to the extent that railroads' finances allowed.

Controlling the temperature expansion of long welded rails proved not so difficult as first thought. It was found that the problem could be minimized by extensive anchorage of the rails to the sleepers or ties to prevent them from moving when the temperature changes, by the use of a heavy ballast section, and by heating the rails before laying to a temperature close to the mean temperature prevailing in the particular locality.

## Rail Fittings

Whether in standard or long welded lengths, rails are joined to each other and kept in alignment by fishplates or joint bars. The offset-head spike is the least expensive way of fastening the rails to wooden crossties, but several different types of screw spikes and clips are used. The rails may be attached directly to wooden crossties, but except on minor lines it is standard practice to seat the rail in a tie plate that distributes the load over a wider area of the tie. A screw or clip fitting must be used to attach rails to concrete ties. A pad of rubber or other resilient material is always used between the rail and a concrete tie.

## Sleepers (Crossties)

Timber has been used for railroad sleepers or ties almost from the beginning, and it is still the most common material for this purpose. The modern wood sleeper is treated with preservative chemical to improve its life. The cost of wood ties has risen steadily, creating interest in ties of other materials.

Steel ties have been used in certain European, African, and Asian countries. Concrete ties, usually reinforced with steel rods or wires, or ties consisting of concrete blocks joined by steel spacing bars are the popular alternative to wooden ties. A combination of concrete ties and long welded rails produces an exceptionally solid and smooth-riding form of track. Concrete ties have been standardized for the main lines of most European railroads and in Japan. Use of concrete elsewhere is increasing—although in North America, which has no European- or Asian-style high-speed rail and where hardwood for traditional crossties is cheap, there is no widespread use.

## Track Maintenance

Modern machinery enables a small group of workers to maintain a relatively long stretch of railroad track. Machines are available to do all the necessary track maintenance tasks: removing and inserting ties, tamping the ballast, cleaning the ballast, excavation and replacement of worn ballast, spiking rail, tightening bolts, and aligning the track. Some machines are equipped to perform more than one task—for example, ballast tamping combined with track lining and leveling. Mechanized equipment also can renew rail, either in conventional bolted lengths or with long welded lengths; a modern machine of this type has built-in devices to lift and pass the old rail to flatcars at its rear and to bring forward and deposit new rail, so that it dispenses with separate crane vehicles.

Complete sections of track—rails and crossties—may be prefabricated and laid in the track by mechanical means. Rail-grinding machines run over the track to even out irregularities in the rail surface. Track-measurement cars, under their own power or coupled into regular trains, can record all aspects of track alignment and riding quality on moving charts, so that maintenance forces can pinpoint the specific locations needing corrective work. Detector cars move over the main-line tracks at intervals with electronic-inspection apparatus to locate any internal flaws in the rails.

The mechanization of track maintenance after World War II has constituted a technologic revolution comparable to the development of the diesel locomotive and electrification. Precision of operation, especially in maintenance of true track alignment, has gained much from the application of electronics to the machines' measuring and control devices. In Europe in particular, highly sophisticated maintenance machines have come into use.

## Auxiliary Plant

Railroad fixed plant consists of much more than the track. More than two-thirds of Germany's new Hannover-Würzburg high-speed line, for example, is in one of its tunnels or bridges or in cutting (excavations). Railroad civil-engineering forces also are concerned with constructing and maintaining thousands of buildings, ranging from small sheds to huge passenger terminals.

## Bridges

The designer of a railroad bridge must allow for forces that result from the concentrated impact that occurs as a train moves onto the bridge; the pounding of wheels, the sidesway of the train, and the drag or push effect as a train is braked or started on a bridge. These factors mean that a railroad bridge must be of heavier construction than a highway bridge of equal length.

As axle loadings become heavier and train speeds higher, bridges need to be further strengthened. Another major objective in modern railroad-bridge construction is the need to minimize maintenance costs. The use of weathering steel, which needs no painting, all-welded construction, and permanent walkways for maintenance personnel contribute to this end. In the advanced countries there has been a widespread trend toward reinforced concrete structures.

## Buildings

Railroad buildings have become fewer and more functional. With paved highways running almost everywhere in the developed countries, it has become more economical to concentrate both freight and passenger operations at fewer stations that are strategically sited and have good highway access. Provision for intermodal traffic exchange has become increasingly important. Particularly in conurbations, the forecourt and surroundings of new passenger stations are laid out to provide adequate and convenient areas for connecting bus or trolley-car services, for private automobile parking, or for so-called "kiss-and-ride"—automobiles that are discharging or picking up rail passengers. Many existing stations have had their surroundings reorganized to provide these facilities.

Many new local stations have been built to serve the spread of commuter and rapid-transit rail systems. However, except on high-speed intercity lines, or at some airports, few sizable city stations have been newly constructed. On the other hand, there has been major reconstruction, updating, and expansion of facilities within the historic fabric of many major city stations in western Europe and in Asia. Particularly in Germany one objective of this rebuilding has been to create easy interchange between ground-level platforms and new metro line platforms below ground. Reconstructed German city stations are also unparalleled for their range of shopping, snack-bar, and restaurant facilities. Another reason for reconstruction has been special provision for new high-speed train services; examples are the Atocha, Nord, and Waterloo termini in Madrid, Paris, and London, respectively. The majority of stations built to serve city airports, generally from platforms beneath a main airport terminal, are on branches of a city's commuter rail system. Those at Frankfurt (Germany), Schiphol (Netherlands), Gatwick (England), and Zurich and Geneva (Switzerland) are directly connected to their national railroad's intercity passenger services.

Diesel and electric locomotives require few maintenance shops as compared with steam locomotives. Car shops, too, have been reduced in number and made more efficient through the use of process-line techniques. It is usually more efficient to construct new shop buildings rather than convert old ones to handle modern types of rolling stock.

## Tunnels

Although very expensive, tunneling provides the most economical means for railroads to traverse mountainous terrain, to gain access to the heart of a crowded city, or, more recently in Japan and Europe, to project a railway across a maritime strait below its seabed. Railroad tunnels, however, confront the construction engineer with some unique problems, particularly in the ventilation of very long bores and in mastery of difficult geologic conditions.

## Railroad Operations and Control

Because a railroad's factory—its plant and train operations—may be spread out over thousands of miles and hundreds of communities, and because its trains use fixed tracks, unlike automobiles or airplanes, it has operating and service problems in some respects more complex than those of a major manufacturing installation. It is not surprising, therefore, that railroads have been among the pioneers in the use of improved methods of communication and control, from the telegraph to the computer and automation techniques.

## Communications

Railroads were among the first to adopt the electric telegraph and the telephone, both for dispatching trains and for handling other business messages. Today, the railroads are among the larger operators of electronic communications systems.

## Radio

Railroads began experimenting with radio at a very early date, but it became practical to use train radio on a large scale only after World War II, when compact and reliable very-high-frequency two-way equipment was developed. In train operations radio permits communication between the front and rear of a long train, between two trains, and between trains and ground traffic controllers. It also is the medium for automatic transmission to ground staff of data generated by the microprocessor-based diagnostic equipment of modern traction and train-sets.

In terminals two-way radio greatly speeds yard-switching work. Through its use, widely separated elements of mechanized track-maintenance gangs can maintain contact with each other and with oncoming trains. Supervisory personnel often use radio in automobiles to maintain contact with the operations under their control.

As the demand for more railroad communication lines has grown, the traditional lineside telegraph wire system has been superseded. As early as 1959, the Pacific Great Eastern Railway in western Canada began to use microwave radio for all communications, doing away almost entirely with line wires. Other railroads all over the world turned to microwave in the 1970s and '80s. More recently many railroads have adopted optical-fibre transmission systems. The high-capacity optical-fibre cable, lightweight and immune to electromagnetic interference, can integrate voice, data, and video channels in one system.

## Computers

A major reason for the growing use of microwave and optical-fibre systems was the tremendously increased demand for circuits that developed from the railroads' widespread use of electronic computers.

Earlier, railroads had been among the leaders in adopting punched-card and other advanced techniques of data processing. In the 1970s and '80s there was a strong trend toward "total information" systems built around the computer. In rail freight operation, each field reporting point, usually a freight-yard office or terminal, is equipped with a computer input device. Through this device, full information about every car movement (or other action) taking place at that point can be placed directly into the central computer, usually located at company headquarters. From data received from all the field reporting points on the railroad, the computer can be programmed to produce a variety of outputs. These include train-consist reports (listing cars) for the terminal next ahead of a train, car-location reports for the railroad's customer-service offices, car-movement information for the car-records department, revenue information for the accounting department, plus traffic-flow data and commodity statistics useful in market research and data on the freightcar needs at each location to aid in distributing empty cars for loading. Tracing of individual car movements can be elaborated by adoption of automatic car identification systems, in which each vehicle is fitted with an individually coded transponder that is read by strategically located electronic scanners at trackside. Major customers can be equipped for direct access to the railroad computer system, so that they can instantly monitor the status of their freight consignments. Relation of real-time inputs to nonvariable data banked in computer memory enables the railroad's central computer to generate customer invoices automatically. Data banks can be developed to identify the optimal routing and equipment required for specific freight between given terminals, so that price quotations for new business can be swiftly computer-generated.

Computers and microprocessors have found many other uses as a railroad management aid. For example, daily data on each locomotive's mileage and any special attention it has needed can be fed by its operating depot into a central computer banking historical data on every locomotive operated by the railroad. In the past, many railroads scheduled locomotive overhauls at arbitrarily assessed intervals, but use of a computer base enables overhaul of an individual locomotive to be precisely related to need, so that it is not unnecessarily withdrawn from traffic. The same procedure can be applied to passenger cars. Systems have been developed that optimize economical use of locomotives by integrated analysis of traffic trends, the real-time location of locomotives, and the railroad's route characteristics to generate the ideal assignment of each locomotive from day to day.

Computerization has given a railroad's managers a complete, up-to-the-minute picture of almost every phase of its operations. Such complete information and control systems have proved a powerful tool for optimizing railroad operations, controlling costs, and producing better service.

## Signaling

Railroad signals are a form of communication designed to inform the train crew, particularly the engine crew, of track conditions ahead and to tell it how to operate the train.

Methods of controlling train operations evolved over many years of trial and error. A common method in the early years was to run trains on a time-interval system; i.e., a train was required to leave a station a certain number of minutes behind an earlier train moving in the same direction. The development of distance-interval systems was a great improvement. In these so-called block systems, a train is prevented from entering a specific section of track until the train already in that section has left it.

Operation of single-track routes on the basis of a timetable alone, which was common on early lines in the United States, had the disadvantage that, if one train were delayed, others also would be delayed, since it was impossible to change the meeting points. By using the telegraph, and later the telephone, the dispatcher could issue orders to keep trains moving in unusual circumstances or to operate extra trains as required. This "timetable–train order" system is still used on many lines in the United States and Canada as well as in developing countries. It is often supplemented with automatic block signals to provide an additional safety factor, and radio is increasingly the means of communication between dispatchers and train crews.

## Types of Signals

The earliest form of railroad signal was simply a flag by day or a lamp at night. The first movable signal was a revolving board, introduced in the 1830s, followed in 1841 by the semaphore signal. One early type of American signal consisted of a large ball that was hoisted to the top of a pole to inform the engineman that he might proceed (hence, the origin of the term highball).

The semaphore signal was nearly universal until the early years of the 20th century, when it began to be superseded by the colour-light signal, which uses powerful electric lights to display its aspects. These are usually red, green, and yellow, either singly or in simultaneous display of two colours. The different colours are obtained either by rotating appropriate roundels or colour filters in front of a single beam or by providing separate bulbs and lenses for each colour. The number of lights and the range of aspects available from one signal can vary depending on its purpose. For instance, additional lights may be installed to the left or right of the main lights to warn a driver of divergence ahead from the through track. In Britain suitably angled strips of white lights are added to signals and illuminated when a divergent track is signaled. Red (stop or danger), green (track clear), and yellow (warning) have the same basic significance worldwide, but in Europe particularly they also are used in combinations of two colours to convey meanings that can vary from one railroad to another. Colour-light signaling is now standard on all but some minor rural lines of the world's principal railways, and its use is spreading elsewhere.

## Automated Systems

The basis of much of today's railroad signaling is the automatic block system, introduced in 1872 and one of the first examples of automation. It uses track circuits that are short-circuited by the wheels and axles of a train, putting the signals to the rear of the train, and to the front as well on single track, at the danger aspect. A track circuit is made by the two rails of a section of track, insulated at their ends. Electric current, fed into the section at one end, flows through a relay at the opposite end. The wheels of the train will then short-circuit the current supply and de-energize the relay.

In a conventional automatic block system, permissible headway between trains is determined by the fixed length of each block system and is therefore invariable. Modern electronics has made possible a so-called "moving block" system, in which block length is determined not by fixed ground distance but by the relative speeds and distance from each other of successive trains. In a typical moving block system, track devices transmit to receivers on each train continuous coded data on the status of trains ahead. Apparatus on a train compares this data with the train's own location and speed, projects a safe stopping distance ahead, and continuously calculates maximum speed for maintenance of that headway. Moving block has been devised essentially for urban rapid-transit rail systems with heavy peak-hour traffic and on which maximum train speeds are not high; in such applications its flexibility by comparison with fixed block increases the possible throughput of trains over one track in a given period of time.

To ensure observance of restrictive signals, a basic form of automatic train control has been used by many major railroads since the 1920s. When a signal aspect is restrictive, an electromagnetic device is activated between the rails, which in turn causes an audible warning to sound in the cab of any train passing over it. If the operator fails to respond appropriately, after a short interval the train brakes are applied automatically. A refinement, generally known as automatic train protection (ATP), has been developed since World War II to provide continuous control of train speed. It has been applied principally to busy urban commuter and rapid-transit routes and to European and Japanese intercity high-speed routes. A display in the cab reproduces either the aspects of signals ahead or up to 10 different instructions of speed to be maintained, decelerated to, or accelerated to, according to the state of the track ahead. Failure to respond to a restrictive instruction automatically initiates both power reduction and braking. The cab displays are activated by on-train processing of coded impulses passed through either the running rails or track-mounted cable loops and picked up by inductive coils on the train. On some high-speed passenger lines the ATP system obviates use of traditional trackside signals.

Among other automatic aids to railroad operation is the infrared "hotbox detector," which, located at trackside, detects the presence of an overheated wheel bearing and alerts the train crew. The modern hotbox detector identifies the location in the train of the overheating and, employing synthesized voice recording, radios the details to the train crew. Broken flange detectors are used in major terminals to indicate the presence of damaged wheels. Dragging equipment detectors warn crews if a car's brake rigging or other component is dragging on the track.

## Interlocking and Routing

The first attempts at interlocking switches and signals were made in France in 1855 and in Britain in 1856. Interlocking at crossings and junctions prevents the displaying of a clear signal for one route when clearance has already been given to a train on a conflicting route. Route-setting or route-interlocking systems are modern extensions of this principle. With them the signaling operator or dispatcher can set up a complete route through a complicated track area by simply pushing buttons on a control panel. Most interlockings employ electrical relays, but adoption of computer-based solid-state interlocking began in Europe and Japan in the 1980s. Safeguard against malfunction is obtained by duplication or triplication; parallel computer systems are arranged to examine electronic route-setting commands in different ways, and only if automatic comparison shows no discrepancy in their proof that conflicting routes have been secured will the apparatus set the required route.

Electronics have greatly widened the scope for precise but at the same time labour-saving control of a busy railroad's traffic by making it possible to oversee extensive areas from one signaling or dispatching centre. This development is widely known as centralized traffic control (CTC). In Britain, for example, one signaling centre can cover more than 320 km (200 miles) of route with a principal city at the hub; the layout under control—used by intercity passenger, suburban passenger, and freight trains—may include 450 switch points and 1,200 possible route-settings. In the United States, the Union Pacific Railroad Company has consolidated dispatching control of its entire system in a single centre at its Omaha, Neb., headquarters.This concentration of signal and point control is possible because of the electronic ability to convey over a single communications channel a multitude of split-second, individually coded commands to ground apparatus and to return confirmations of compliance equally rapidly.

The functions of track circuits have been multiplied by electronics. The individual timetable number or alpha-numeric code of a train is entered into the signaling system at the track-circuited block where the train starts its journey. As the train moves from one block section to another, its occupation of successive track circuits automatically causes its number or code to move accordingly from one miniature illuminated window to another on the signaling centre's layout displays. When the train moves from one control area to another, its code will automatically move to the next centre's layout display. The real-time data on individual train progress generated by this system can be adapted for transmission to any interested railway office or, on a passenger railroad, to drive service information displays at stations. Particularly on rapid-transit systems, setting of junctions can be automated if train numbers or codes include an indication of routing, which is electronically detected when they occupy a track circuit at the approach to the divergence.

From the foregoing it is apparent that the means for complete automation of train operation exist. It has been applied to some private industrial rail systems since the early 1970s, and most of the capability has been built into some city metro systems. Extension of computer processing to the real-time data on train movement generated from track circuitry has further benefited control of major railroads' traffic. In Europe's latest centres controlling intensive passenger operations, operators can call up graphic video comparisons of actual train performance with schedule, projections of likely conflict at junctions where trains are not running on schedule, and recommendations for revision of train priorities to minimize disruption of scheduled operation. In North America, where many main lines are single-track, the Computer-Assisted Dispatching System (CADS) can relieve the operator of much routine work. At Union Pacific's Omaha centre, once the dispatcher has entered a train's identity and priority, the system automatically routes it accordingly, arranging its passing of other trains in loops as befits its priority. CADS automatically updates and modifies its determinations based on actual train movements and changing track conditions. The operator can intervene and override the system.

In early CTC installations the layout under a centre's control was shown only on one panoramic display, in which appropriately located lights indicated the setting of each switch point and signal, the track-circuited sections occupied by trains, and in windows at each occupied section the identifying code of the train in question. In some installations route-setting buttons were incorporated in this display. In the most recent CTC centres the overall panoramic display is generally retained, but operators have colour video screens portraying close-ups of the areas under their specific control. In many such cases, a light-pencil or tracker-ball movement of a cursor is used to identify on the screen the route to be changed. Alternatively, the operators may have alphanumeric keyboards on which reset route codes may be entered.

On the main lines of North America, precise control of train movement is more difficult than in Europe, because block sections are much longer. To overcome the problem, the principal railroads of the United States and Canada combined in the 1980s to develop an Advanced Train Control Systems (ATCS) project, which integrated the potential of the latest microelectronics and communications technologies. In fully realized ATCS, trains continuously and automatically radio to the dispatching centre their exact location and speed; both would be determined by a locomotive-mounted scanner as well as signals received from global positioning system (GPS) satellites. In the dispatching centre, this input is processed to arrive at the optimal speed for each train in relation to its priority, the proximity of other trains it must pass, and route characteristics. From this analysis, continuously updated instructions can be radio-transmitted to train locomotives and processed by onboard computers for reproduction on cab displays so that trains can be driven with maximum regard for operating and fuel-consumption efficiency. ATCS can be developed in several stages, or levels, up to full implementation.

### The Marshaling Yard

A major area for automation techniques in railroading is the large classification, or marshaling, yard. In such yards, freight cars from many different origins are sorted out and placed in new trains going to the appropriate destinations. Marshaling yards are frequently called "hump yards" because the large installations have a "hump" over which cars are pushed. The cars then roll down from the hump by gravity, and each is routed into a classification or "bowl" track corresponding to its destination or where the train for the next stage of its transit is being formed.

Operations in classification yards have reached a high degree of automation. The heart of the yard is a central computer, into which is fed information concerning all cars in the yard or en route to it. As the cars are pushed up the hump (in some yards, by locomotives that are crewless and under remote radio control from the yard's operations centre), electronic scanners confirm their identity by means of a light-reflective label, place the data (car owner, number, and type) in a computer, and then set switches to direct each car into the proper bowl track. Electronic speed-control equipment measures such factors as the weight, speed, and rolling friction of each car and operates electric or electropneumatic "retarders" to control the speed of each car as it rolls down from the hump. Every phase of the yard's operations is monitored by a computerized management control and information system. With hand-held computers, ground staff can input data directly into the yard's central computer.

Because such electronically equipped yards can sort cars with great efficiency, they eliminate the need to do such work at other, smaller yards. Thus, one large electronic yard usually permits the closing or curtailing of a dozen or more other yards. Most modern electronic yards have quickly paid for themselves out of operating savings—and this takes no account of the benefits of improved service to shippers.

## Ladder Track

Ladder track is a type of railway track in which the track is laid on longitudinal supports with transverse connectors holding the two rails at the correct gauge distance. Modern ladder track can be considered a development of baulk road, which supported rails on longitudinal wooden sleepers. Synonyms include longitudinal beam track.

Ladder type track has also be used historically on bridges lacking ballast, and in situations requiring good drainage or ease of maintenance such as stations.

Described as track with longitudinal bearers with cross ties, ladder track type was common on early British railways (c.1830/40s), including the Birmingham and Gloucester Railway, Dublin and Kingstown Railway, Glasgow, Paisley, Kilmarnock and Ayr Railway, Great Western Railway, Ulster Railway, Newcastle and North Shields Railway, Slamannan Railway (part of), Hull and Selby Railway (two thirds of), Manchester and Bolton Railway (including lontudinal stone sleepers), and the London and Greenwich Railway (partly). On the Hull and Selby Railway it was used in part as it was noted to produce smooth running, and low wheel wear. However the contact between rail and sleeper produced hydraulic pumping in wet conditions, which led to rolling stock becoming dirtied very quickly. The longitudinal track was also found cause issue with wheel slip on inclines, and to be noticeably inferior to transverse sleepered track in terms of traction. No longitudinally laid track remained on the line after 1860.

Research into longitudinal sleepers took place in Japan, Russia and France in the mid 20th century. In the late 20th century, interest in ladder type tracks increased due to its potential for lower cost and lower maintenance railways, as well as increased stability benefits over sleepered track.

In general, ladder track produces lower pressures on the road bed: both the maximum pressure and the amplitude of pressure pulses are smaller than transverse sleepered track, which can reduce maintenance costs in ballasted systems. An additional benefit in ballasted ladder track is increased resistance to ballast wash out and other forms of ballast degradation due the addition longitudinal support and rigidity in ladder track.; the same structural rigidity also adds to buckling resistance.

## Design Variants

### Tubular Modular Track

Tubular Modular Track.

Tubular Modular Track (or TMT) is a type of ballastless ladder track manufactured by Tubular Track (Pty.) Ltd. of South Africa first introduced in 1989. and invented by Peter Küsel.

The track consists of steel rail resting on concrete supports via rubberised cork absorption pads with galvanised steel tie beams which wrap around the concrete members rather than being cast into the concrete as with other ladder tracks. The track is modular and precast, rather than being cast in situ. Modular turnouts are also produced. The modular nature and controlled production of the track sections has the advantage of rapid installation and good quality control. The ballastless and modular nature of the track makes it advantageous for both wet and desert conditions where ballast degradation is problematic, as well as in mines where transportation of the track components is simplified. Since the rail is continuously supported rail stresses are less; a 34 tonne axle load version has been designed and lighter rails can be used compared to sleepered track, additionally maintenance costs of ballast are reduced.

The track has been used mainly in southern Africa, including a section of the Gautrain line in South Africa. The system has also been used in Saudi Arabia.

### RTRI Japan Ladder Tracks

Ladder track.

The Railway Technical Research Institute of Japan has developed two types of ladder track; ballasted and a floating un-ballasted type.

Both types utilise a 'combined rail' of steel rails each attached through a conventional modern railway fastener to a longitudinal pre-stressed concrete supports (6.25m long) which are kept in gauge by thick walled tubular steel cross members embedded in the concrete. Forms for axle load of 40 tonnes have been designed.

The ballasted track show reduced maintenance (tamping) compared to conventional cross-tie sleeper track, longitudintal creep is greater than that in conventional track; for this reason transverse 'anti creep' panels are included in the structure.

The non-ballasted track is supported by vibration absorbing springs (or elastomeric pads) on a concrete road bed; the combined structure shows reduced vertical vibrations of the track bed compared to conventional track This property gives reduced noise levels when the track is utilised in a steel girder bridge.

### Other Types

Specialised Track Systems (Pty.) Ltd. licenses its technology for ladder track to contractors; the track system is ballastless with steel lateral braces in a concrete roadbed, support of the track is by conventional rail fastening systems spaced at regular intervals. The design can also incorporate ducts within the beams (for cabling) and can be converted to slab track by in-situ concrete pouring. The company's main market is mining applications.

## Ballastless Track

A ballastless track or slab track is a type of railway track infrastructure in which the traditional elastic combination of ties/sleepers and ballast is replaced by a rigid construction of concrete or asphalt.

## Characteristics

In ballastless tracks, the rails are rigidly fastened to a special type of concrete ties/sleepers that are themselves set in concrete. Ballastless tracks therefore offer a high consistency in track geometry, the adjusting of which is not possible after the concreting of the superstructure. Therefore, ballastless tracks must be concreted within a tolerance of 0.5 millimetres (0.0197 in). The elasticity of the ballast in the traditional railway superstructure is replaced by flexibility between either the rails and the concrete ties/sleepers or the ties/sleepers and the concrete or asphalt slab as well inherent elasticity within the conglomerate of the tie/sleeper, whereas the concrete or asphalt slab is usually inelastic.

## Advantages

The advantages of a ballastless track over a traditional superstructure are its highly consistent track geometry, its longer life span, and the reduced need for maintenance.

A ballastless track's track geometry is achieved mainly due its relative inelasticity in comparison to a traditional superstructure that results in far fewer deformations and generally smoother running; drivers of the London Overground's East London Line have informally declared the Low Vibration Track system the smoothest superstructure they have experienced. Measurements conducted in Switzerland in 2003 and 2004 showed a standard deviation of the gauge of less than 1.2 millimetres (0.047 in).

This in turn increases the track's life span and reduces the need for maintenance. The usual preventive maintenance is limited to rail grinding, since tamping is not necessary due to the absence of ballast. Curative maintenance beyond rail replacement is required only after several decades. The Swiss Federal Railways replaced the ties/sleepers and rubber shoes of the ballastless track in the 4.9-kilometre (3.0 mi)-long Heitersberg tunnel between 2014 and 2016, whereas no maintenance of the concrete slab was necessary 39 years after the tunnel's opening. Due to its good experiences with the system, the Swiss Federal Railways are looking to install ballastless track wherever there is a rigid substructure—in tunnels as well as on viaducts.

Further advantages of ballastless tracks include better and controlled drainage, the elimination of flying-ballast damage on rolling stock and civil engineering structures, a shallower superstructure, and the possibility run-over sections such as crossings over which pneumatic vehicles can be driven. When used in stations, ballastless tracks are easier to clean.

## Disadvantages

The primary disadvantage of a ballastless track is its significantly higher cost of initial construction. While numbers vary depending on construction type and track infrastructure (ballastless tracks are generally more suitable to infrastructures that are also made of concrete, as is the case in tunnels or on viaducts), the Deutsche Bahn estimated in 2015 that construction costs of ballastless tracks are 40 percent higher than those of traditional superstructure. However, the life-cycle cost of ballastless tracks are generally lower than those of ballasted tracks due to significantly lower maintenance.

Further disadvantages of ballastless tracks are the impossibility of adjusting or correcting track geometry once concrete has been set, the necessity of a stable infrastructure (since no adjustments can be made to the superstructure), higher noise emissions, and longer repair times when the concrete slab is damaged (e.g. due to construction faults or wear and tear).

## Construction Types

The following list contains construction types of ballastless tracks that have been internationally used in heavy-rail systems (as opposed to light railways, tramways, or metros) in chronological order of their first usage.

### SBB Bözberg/STEDEF (SBB)

The Bözberg/STEDEF system consists of twin ties/sleepers that are connected by a steel track rod and enclosed in a rubber shoe. All of its components can be exchanged individually. Bözberg/ STEDEF was first used by the Swiss Federal Railways in the Bözberg tunnel in 1966. STEDEF was further developed by SATEBA prior to the system's installation on the French LGV Méditeranée.

Ballastless track of the type "Rheda 2000" between rail tunnels under the Dordtsche Kil.

### Rheda (Rail.One)

The Rheda system consists of three layers: a base course and two slabs that are joined by rebars, as are the individual ties/sleepers. Rheda was first used by the Deutsche Bahn in Rheda-Wiedenbrück station, after which it is named, in 1972. It has since been installed on the Dutch HSL-Zuid route between Amsterdam and Rotterdam, in the Spanish Guadarrama and Sant Joan Despí tunnels, and on the Chinese Wuhan–Guangzhou High-Speed Railway.

### Bögl (Max Bögl)

The Bögl ballastless track is characterised by its use of prefabricated concrete slabs in lieu of a continuous structure that is cast on site. Mortar is used to connect the 9-ton-slabs to the infrastructure and to one another. The Bögl system was developed in Germany and first tested in Dachau in 1977. The first serial installation took place in Schleswig-Holstein and Heidelberg in 1999. For its use on the high-speed link between Beijing and Shanghai, 406,000 slabs were installed.

### FF ÖBB/PORR (PORR)

The ÖBB/PORR ballastless track (FF stands for German *Feste Fahrbahn*, meaning *ballastless* or, literally, *fixed track*) consists of an elastically supported track slab. It was first tested in 1989, became the standard system in Austria in 1995, and has been used for over 700 kilometres of track worldwide, including the German Verkehrsprojekt Deutsche Einheit Nr. 8 and the Doha metro.

Ballastless track of the type "Low Vibration Track" in the Gotthard Base Tunnel's Sedrun multifunction station.

## Low Vibration Track (Sonneville/Vigier Rail)

The Low Vibration Track (LVT) system is similar to Bözberg/STEDEF in that it also uses twin ties/sleepers enclosed in rubber shoes. However, LVT does not feature a tie rod. The system was developed and tested by Roger Sonneville together with the Swiss Federal Railways in the 1990s before the rights were sold to Vigier Rail in 2009. LVT has been in service in the Channel tunnel since 1994. Due to the tunnel's German name *Eurotunnel*, LVT is sometimes referred to as Euroblock. LVT has been used for over 1300 kilometres of track worldwide, including the Swiss Lötschberg, Gotthard and Ceneri base tunnels, the South Korean high-speed Suin Line between Songdo and Incheon, the Turkish Marmaray project, and the London Overgound's East London line, as well as on viaducts in urban areas. LVT has become the standard ballastless-track system in Switzerland.

## IVES

The IVES system (*Intelligent, Versatile, Efficient and Solid*) is a product of Rhomberg Rail. The system consists of a base layer (preferably common asphalt concrete) and concrete lateral structural elements, in which the rail fastening elements of the type DFF 304 are directly embedded – no ties/sleepers are needed. The necessary elasticity is given only by a flexible intermediate plate in the rail fastening elements.

The structural elements of this system are individually manufactured and can be positioned laterally or longitudinally onto the base layer. The structural elements have recesses in the top, where the rail fastening elements are placed into. Afterwards, the rails are lifted onto the fastening elements and a track grid is established. The grid's exact position can now be adjusted vertically and laterally. At last, the rail fastening elements are friction-locked to the structural elements with high-strength grouting mortar. Thanks to its versatile construction and easy installation, IVES is suitable for all rail types.

After testing, the first IVES track has been installed in the Asfordby Tunnel of the Old Dalby Test Track in Great Britain in 2013 and since then, 7 more IVES tracks have been built. The longest IVES track runs through the Bruggwaldtunnel in Switzerland, with a total length of 1.731 m.

# Structure of Railway Track

The track on a railway or railroad, also known as the permanent way, is the structure consisting of the rails, fasteners, railroad ties (sleepers, British English) and ballast (or slab track), plus the underlying subgrade. It enables trains to move by providing a dependable surface for their wheels to roll upon. For clarity it is often referred to as railway track (British English and UIC terminology) or railroad track (predominantly in the United States). Tracks where electric trains or electric trams run are equipped with an electrification system such as an overhead electrical power line or an additional electrified rail.

The term *permanent way* also refers to the track in addition to lineside structures such as fences.

## Structure

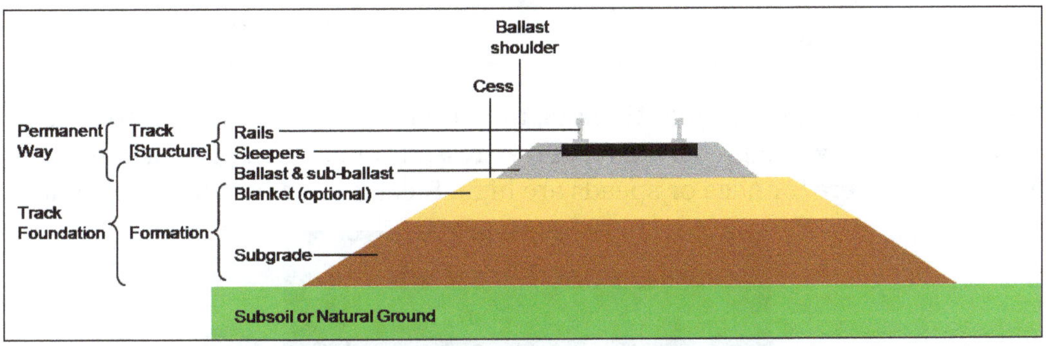


Section through railway track and foundation showing the ballast and formation layers. The layers are slightly sloped to help drainage. Sometimes there is a layer of rubber matting (not shown) to improve drainage, and to dampen sound and vibration.

## Traditional Track Structure

Notwithstanding modern technical developments, the overwhelmingly dominant track form worldwide consists of flat-bottom steel rails supported on timber or pre-stressed concrete sleepers, which are themselves laid on crushed stone ballast.

Most railroads with heavy traffic utilize continuously welded rails supported by sleepers attached via base plates that spread the load. A plastic or rubber pad is usually placed between the rail and the tie plate where concrete sleepers are used. The rail is usually held down to the sleeper with resilient fastenings, although cut spikes are widely used in North American practice. For much of the 20th century, rail track used softwood timber sleepers and jointed rails, and a considerable extent of this track type remains on secondary and tertiary routes. The rails were typically of flat bottom section fastened to the sleepers with dog spikes through a flat tie plate in North America and Australia, and typically of bullhead section carried in cast iron chairs in British and Irish practice. The London, Midland and Scottish Railway pioneered the conversion to flat-bottomed rail and the supposed advantage of bullhead rail - that the rail could be turned over and re-used when the top surface had become worn - turned out to be unworkable in practice because the underside was usually ruined by fretting from the chairs.

Jointed rails were used at first because contemporary technology did not offer any alternative.

However, the intrinsic weakness in resisting vertical loading results in the ballast becoming depressed and a heavy maintenance workload is imposed to prevent unacceptable geometrical defects at the joints. The joints also needed to be lubricated, and wear at the fishplate (joint bar) mating surfaces needed to be rectified by shimming. For this reason jointed track is not financially appropriate for heavily operated railroads.

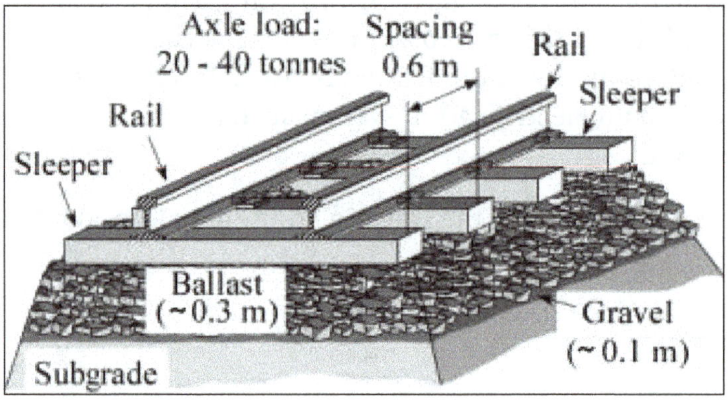


Timber sleepers are of many available timbers, and are often treated with creosote, Chromated copper arsenate, or other wood preservatives. Pre-stressed concrete sleepers are often used where timber is scarce and where tonnage or speeds are high. Steel is used in some applications.

The track ballast is customarily crushed stone, and the purpose of this is to support the sleepers and allow some adjustment of their position, while allowing free drainage.

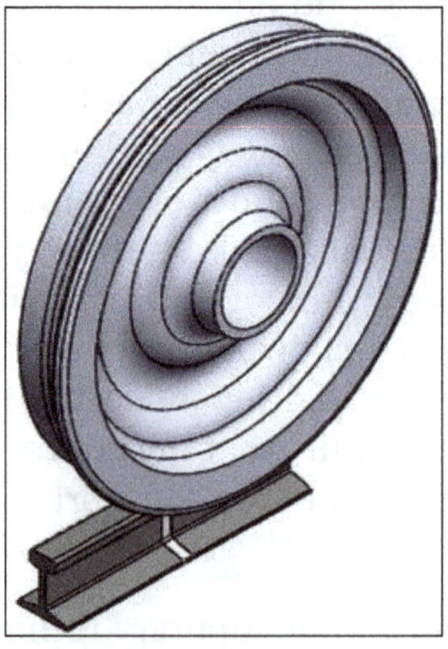

## Requirement of Track Material

If the track is an ST sleeper track with 52-kg rails and sleeper density M + 7 (where M is the length of the rail in metres, which is normally 13 m on BG), then the calculations are as follows:

$$\text{Number of rails per km} = \frac{1000}{M} \times 2 = \frac{1000}{13} \times 2 = 76.9 \times 2$$
$$= \text{Approx.} 154$$
$$\text{Weight of rails per km} = 154 \times 52\text{kg}$$
$$= 8008\text{kg} = 8.08\text{mt}$$

Sleepers:

$$\text{Number of sleeper per km} = (M+7) \times \frac{\text{total number of rails}}{2}$$
$$= (13+7) \times \frac{154}{2} = 1540$$

Fish bolts Each joint requires four fish bolts:

Number of fish bolts = 4 x number of rails per km = 4 x 154 = 516.

Fish plates Each joint requires two fish plates:

Number of fish plates = 2 x number of rails per km = 2 x 154 = 308.

Keys Four keys are required per sleeper:

Number of keys = 4 x 1540 = 5160.

Loose jaws Four loose jaws are required per sleeper:

Number of loose jaws = 1540 x 4 = 5160.

Track renewals are required every year to keep the track safe and in good running condition. The tracks on Indian Railways require heavy renewal since they are very old and have been in use for a very long time. Track renewal is done to rectify defects in rails, fish plates, and sleepers. The track renewal train is used on Indian Railways for mechanized renewal of the track. It can lay 16 sleepers per minute and is able to give an average output of 200-250 km/year.

## Track Geometry

Track geometry is three-dimensional geometry of track layouts and associated measurements used in design, construction and maintenance of railroad tracks. The subject is used in the context of standards, speed limits and other regulations in the areas of track gauge, alignment, elevation, curvature and track surface. Although, the geometry of the tracks is three-dimensional by nature, the standards are usually expressed in two separate layouts for horizontal and vertical.

# Layout

## Horizontal layout

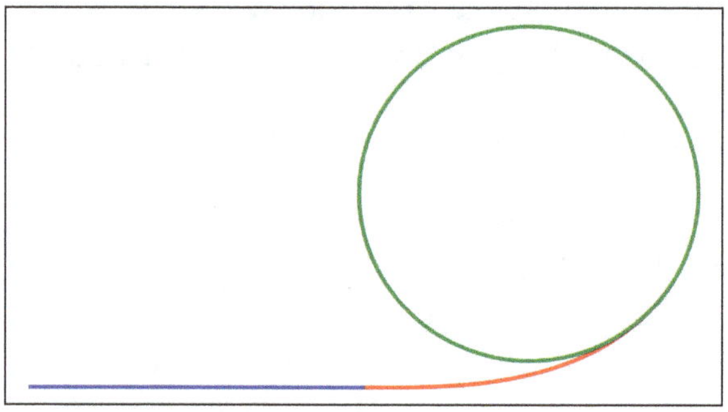

Tangent track in blue with transition spiral in red and curved track in green.

Horizontal layout is the track layout on the horizontal plane. This can be thought of as the plan view which is a view of a 3-dimensional track from the position above the track. In track geometry, the horizontal layout involves the layout of three main track types: tangent track (straight line), curved track, and track transition curve (also called transition spiral or spiral) which connects between a tangent and a curved track.

In Australia, there is a special definition for a bend (or a horizontal bend) which is a connection between two tangent tracks at almost 180 degrees (with deviation not more than 1 degree 50 minutes) without an intermediate curve. There is a set of speed limits for the bends separately from normal tangent track.

## Vertical Layout

Vertical layout is the track layout on the vertical plane. This can be thought of as the elevation view which is the side view of the track to show track elevation. In track geometry, the vertical layout involves concepts such as crosslevel, cant and gradient.

## Reference Rail

The reference rail is the base rail that is used as a reference point for the measurement. It can vary in different countries. Most countries use one of the rails as the reference rail. For example, the United States uses the reference rail as the line rail which is the east rail of tangent track running north and south, the north rail of tangent track running east and west, the outer rail (the rail that is further away from the center) on curves, or the outside rails in multiple track territory. For Swiss railroad, the reference rail for tangent track is the center line between two rails, but it is the outside rail for curved track.

## Track Gauge

Track gauge or rail gauge (also known as track gage in the United States) is the distance between the inner sides (gauge sides) of the heads of the two load bearing rails that make up a single railway line. Each country uses different gauges for different types of trains. However, the 1,435 mm (4 ft 8 $\frac{1}{2}$ in) gauge is the basis of 60% of the world's railways.

## Transverse Elevation

### Crosslevel

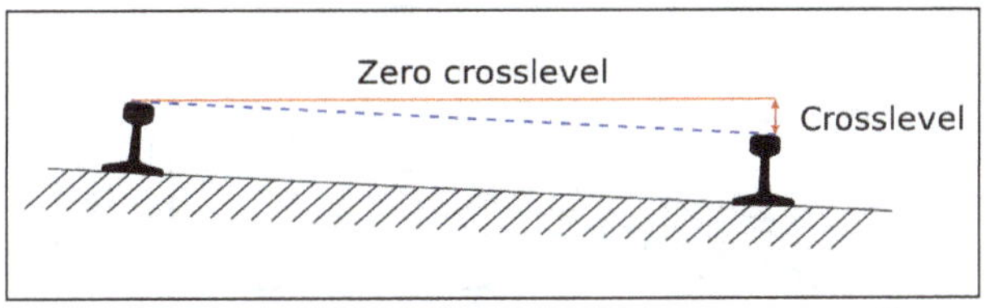


The measurement of crosslevel between two rails.

Crosslevel (or 'cross level') is the measurement of the difference in elevation (height) between the top surface of the two rails at any point of railroad track. The two points (each at the head of each rail) are measured at by the right angles to the reference rail. Since the rail can slightly move up and down, the measurement should be done under load.

It is said to be zero crosslevel when there is no difference in elevation of both rails. It is said to be reverse crosslevel when the outside rail of curved track has lower elevation than the inside rail. Otherwise, the crosslevel is expressed in the unit of height.

The speed limits are governed by the crosslevel of the track. In tangent track, it is desired to have zero crosslevel. However, the deviation from zero can take place. Many regulations have specification related to speed limits of certain segment of the track based on the crosslevel.

For curved track, most countries use the term cant or superelevation to express the difference in elevation and related regulations.

### Warp

Warp is the difference in crosslevel of any two points within the specific distance along the track. The warp parameter in the track geometry is used to specify the maximum in the crosslevel difference of the track in any segment (tangents, curves and spirals).

Without the maximum warp parameter, the regulation on crosslevel alone may not be sufficient. Consider rails with a positive crosslevel followed by a negative crosslevel followed by a sequence of alternating positive and negative crosslevels. Although, all of those crosslevels are in permissible parameter, when operating a train along such track, the motion will be rocking left and right. Therefore, the maximum warp parameter is used to prevent the critical harmonic rock-off condition that may result in the trains rocking back and forth and derailing following wheel climb.

In the United States, the specific distance used for measurement to ensure that the difference in crosslevel of the track is within the permissible warp parameter is 62 feet. The design warp is zero for both tangent and curved track. That means, ideally, the crosslevel should not change between any two points within 62 feet. There are some deviations to allow crosslevels along the track to change (such as change for superelevation in curves). Different levels of those deviations from the zero warp specify the speed limits.

The specification that focuses on the rate of change in crosslevels of curved track is contained within the area related to cant gradient.

## Longitudinal Elevation

### Track Gradient

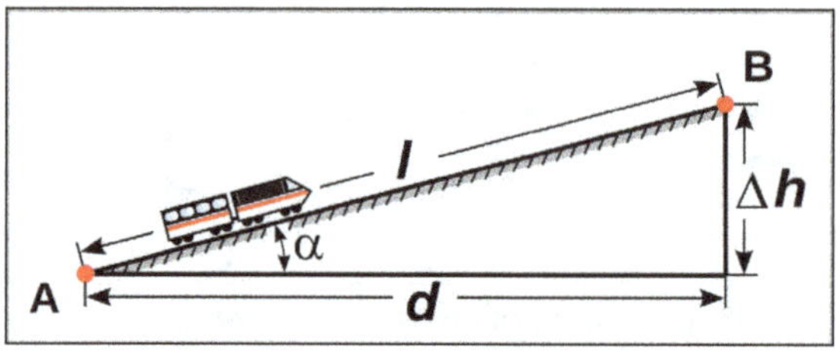


d = distance traveled horizontally, Δh = rise, l = slope length, α = angle of inclination.

The term track gradient is relative elevation of the two rails along the track. This can be expressed in the distance traveled horizontally for a rise of one unit, or in terms of an angle of inclination or a percentage difference in elevation for a given distance of the track.

The allowable gradients may be based on the ruling gradient which is the maximum gradient over which a tonnage train can be hauled with one locomotive. In some countries, momentum gradient which is a steeper but shorter gradient may be allowed. This is usually when there is a track gradient is connected to a leveled tangent track that is long enough with no signal between them so that train can build momentum to push through steeper grade than it can be without momentum.

In curved track (with or without cant), there will be curve resistance to push the trains through the curve. The allowable gradients may be reduced on curves to compensate for the extra curve resistance. The gradient should be uniform along the track.

### Vertical Curve

Vertical curve is the curve in vertical layout to connect two track gradients together whether it is for changing from an upgrade to a downgrade (summit), changing from a downgrade to an upgrade (sag or valley), changing in two levels of upgrades or changing in two levels of downgrades.

Some countries do not have specification on the exact geometry of vertical curves beyond general specification on vertical alignment. Australia has specification that the shape of vertical curves should be based on quadratic parabola but the length of a given vertical curve is calculated based on circular curve.

### Curvature

Curve with superelevation of tracks on the Keystone Corridor near Rosemont, PA.

In most countries, the measurement of curvature of curved track is expressed in radius. The shorter the radius, the sharper the curve is. For sharper curves, the speed limits are lower to prevent an outward horizontal centrifugal force to overturn the trains by directing its weight toward the outside rail. Cant may be used to allow higher speeds over the same curve.

In the United States, the measurement of curvature is expressed in degree of curvature. This is done by having a chord of 100 feet (30.48 m) connecting to two points on an arc of the reference rail, then drawing radii from the center to each of the chord end points. The angle between the radii lines is the degree of curvature. The degree of curvature is inverse of radius. The larger the degree of curvature, the sharper the curve is. Expressing the curve in this way allows surveyors to use estimation and simpler tools in curve measurement. This can be done by using a 62-foot (18.90 m) string line to be a chord to connect the arc at the gage side of the reference rail. Then at the midpoint of the string line (at the 31st foot), a measurement is taken from the string line to the gauge of the reference rail. The number of inches in that measurement is approximated to be the number of degrees of curvature.

Due to the limitation of how specific train equipment can make a turn at maximum speeds, there is a limitation of minimum curve radius to control the sharpness of all curves along a given route. Although most countries use radius for measurement of curvature, the term maximum degree of curvature is still used outside of the United States such as in India, but with the radius as the unit.

## Cant

Railroad track spirit level in place indicating 5" of superelevation between the inside and outside rails of a curve along the Keystone Corridor near Narberth, PA.

In curved track, it is usually designed to raise the outer rail, providing a banked turn, thus allowing trains to maneuver through the curve at higher speeds that would otherwise be not possible if the surface was flat or level. It also helps a train steer around a curve, keeping the wheel flanges from pressing the rails, minimizing friction and wear. The measurement of the difference in elevation between the outer rail and the inner rail is called cant in most countries. Sometime the cant is measured in term of angle instead of height difference. In the United States, it is measured in height difference and called crosslevel, even for the curved track.

When the outside rail is at higher elevation than the inside rail, it is called positive cant. This is normally the desired layout for curved track. Most counties achieve the desired level of positive cant by raising the outside rail to that level which is called superelevation. For Swiss railroads, the cant is done by rotating at the track axis (center of the two rails) to have outside rail super elevated (raised) at the half rate of the desired cant and the inside rail under elevated (lowered) at the same half rate of the desired cant.

When the outside rail is at lower elevation than the inside rail, it is called negative cant (or reverse crosslevel in the United States). This is not usually a desired layout but it may be unavoidable in some situations such as curves involving turnouts.

There are regulations which limit the maximum cant. This is to control the unloading of the wheels on the outside rail (high rail), especially at low speeds.

## Cant Gradient

Cant gradient is the amount by which cant is increased or decreased in a given length of track. The change in cant is required in order to connect a tangent track (no cant) to a curved track (with cant) through a transition curve. The rate of change of cant is used to determine the suitable cant gradient for a given design speed. Track twist may also be used to describe cant gradient which may be expressed in percentage of cant change per length unit. However, in the UK, the term track twist is normally used in the context of cant gradient with higher values which are considered to be faults.

In the United States, the required cant gradient on a transition curve to achieve smooth connection between superelevation of curved track and the zero crosslevel of tangent track is called superelevation runoff. In addition to the runoff specification, the regulations related to allowable rate of change in the cant is also part of the general specification on the rate of change in crosslevel called warp parameter. The warp parameter and superelevation runoff help calculate the required length of the runoff for a transition curve.

## Cant Deficiency

As described, cant may be used to reduce lateral acceleration on trains traveling on curved track. This is to balance the centrifugal force (force pushing outward the curve) and centripetal force (force pushing inward the curve). At a higher speed, the centrifugal force is higher. On the contrary, higher cant creates the higher centripetal force. The calculation for this assumes a constant train speed on a constant radius curve.

When the speed of the train and the amount of cant are in balance (centrifugal matches centripetal), it is called equilibrium. This would make the components of wheel to rail force normal to the plane of the track having the same in aggregate for the outside rail as for the inside rail. This would also make the passengers in the train not to perceive any lateral acceleration (a push toward sideway).

For a fixed amount of cant, the speed that creates balance is called equilibrium speed. For a constant speed of a running train, the amount of required cant to achieve the balance is called equilibrium cant.

In practice, trains are not running on equilibrium cants at curves. The situation is called unbalance, which can be in one of the two following ways. For a given speed, if the actual cant is less than the equilibrium cant, the amount of cant difference is called cant deficiency. In the other word, it is the amount of missing cant to achieve the balance. On the contrary, for a given speed, if the actual cant is higher than the equilibrium cant, the amount of over cant from the balance is called cant excess.

In a shared track configuration for trains with different operating speeds such as freight and higher-speed rail passenger services, the cant on a curve should be considered for both high and low speeds. The higher-speed trains would experience cant deficiency and the lower-speed trains would experience cant excess. These parameters have a significant effect on curve performance which includes safety, passenger comfort, and wear and tear of equipment and rails.

## Alignment

The term alignment is used in both horizontal and vertical layouts to describe the line uniformity (straightness) of the rails.

The horizontal alignment (or alinement in the United States) is done by using a predefined length of string line (such as 62-foot in the US and 20 meters in Australia) to measure along the gauge side of the reference rail. It is the distance (in inches or millimeters) from the midpoint of the string line to the gauge of the reference rail. The design horizontal alignment for tangent track is zero (perfect straight line on the horizontal layout). The design horizontal alignment on the curved track in the United States is 1 inch for each degree of curvature. Any other readings indicate deviations.

The vertical alignment (or profile in the United States, but not to be confused with rail profile) is the surface uniformity in the vertical plane. The measurement of uniformity is done using a predefined length of string line (normally the same length used in horizontal alignment) along the track. If the midpoint of the measurement has higher elevation, it is called hump deviation. On the other hand, if the midpoint has lower elevation, it is called dip deviation.

These deviations from design alignment are used as parameters to assign speed limits.

## Track Spacing

The track spacing is the distance of the track centres of double-track railway lines. There are standard distances derived from the standard loading gauge in a country. For high-speed trains and in tighter curves that distance needs to be increased.

The track spacing is also called the *center-to-center* spacing to differentiate it from the *edge-to-center* spacing of a railway. These two values may be different depending on how signal masts are added to the overall track geometry.

The distance between the track centres makes a difference in cost and performance of a double-track line. The track centres can be as narrow and as cheap as possible, but maintenance must be done on the side. Signals for bi-directional working cannot be mounted between the tracks so must be mounted on the 'wrong' side of the line or on expensive signal bridges. Track centres are usually wider on high-speed lines, as pressure waves knock each other as high-speed trains pass.

The minimum track spacing can be derived from the loading gauge. The European Berne Gauge has a width of 3,150 mm. The minimal distance to structures on the side of the track is half of it but that is doubled again for double track lines. As the wagon can move within the rail gauge one adds 100 mm and with a possible displacement of tracks over time one adds some 250 mm as a security margin. This leads to a minim centre-to-center spacing of 3,500 mm. In fact, the first Prussian railways were built to that standard but it became soon apparent that it was too dangerous with some passengers having a hand or their head out of the window.

Narrow track centres might be 4 metres (13 ft) or less. Narrow track centres may have to be widened on sharp curves to allow for long rail vehicles following the arc of the curve, and this increases a surveyor's workload. Widening a track centre to 5 metres (16 ft) or so suits high-speed trains passing each other, and eliminates the need to widen the centres on sharp curves. Increasing width of track centres of 6 metres (20 ft) or more makes it much easier to mount signals and overhead wiring structures.

In the United Kingdom, the early lines of the Great Western Railway were built to Brunel's broad gauge, which was also associated with a more generous loading gauge. This is still apparent along those lines which remain in use today: structures such as bridges and tunnels are larger, and the distances between opposite platform faces are larger.

Very wide centres at major bridges can have military value. It also makes it harder for rogue ships and barges knocking out both bridges in the same accident.

Railway lines in desert areas affected by sand dunes are sometimes built on alternate routes so that if one is covered by sand, the other(s) are still serviceable.

If the standard track centre is changed, it can take a very long time for most or all tracks to be brought into line.

## Regulations

The general standard in Germany and Switzerland had been to build new tracks with a centre-to-centre spacing of 3.80 m and a spacing of 4.50m in railway stations. Depending on the usage of the tracks it was still possible to build new double track lines with track centres of just 3.50 m. With the trains going faster over time the track centres were increased to 4.00 m on main lines.

The advent of high-speed trains required a stronger regulation which was regulated in Germany's EBO. The first update of 1982 increased the minimum track centre to 4.00 m allowing no more exceptions. But by 1991 it was replaced with a table taking into account the maximum speed of the trains on a track as well as the curvature. A distance of 4.00 m was considered enough for speeds up to 200 km/h. At 250 km/h the tracks have a centre-to-centre distance of 4.50 m (while the first tracks in the 1980s were built with a distance of 4.70 m).

The TGV track construction puts both rail tracks into a common concrete block, so they can disregard a safety margin for track displacement. This allows high-speed rail to have a centre-to-centre distance of just 4.20 m. Additionally, these lines are only allowed for high-speed passenger rail where no out-of-gauge loads are expected and the windows in the trains can not be opened.

In Japan, the first high-speed tracks of the Central Japan Railway Company Shinkansen were built with a distance between track centres of 4.3 m.

The largest minimum track centre is planned for Indias high-speed network requiring a common distance of 5,30 m.

## Trach Centre Examples

- 09 ft 8.0 in (2.95 m) Liverpool and Manchester Railway 1830 at opening day; later widened.

- 10 ft 8.5 in (3.26 m) United Kingdom (a six-foot plus a four foot).
- 11 ft 0.0 in (3.35 m) New South Wales 1855 old standard (estimated).
- 12 ft 0.0 in (3.66 m) New South Wales 1910 new standard for 10 ft 6.0 in (3.20 m) wide carriages. Rounded in imperial.
- 13 ft 1.5 in (4.00 m) New South Wales 1973 metric standard; rounded in metric.
- 14 ft 9.0 in (4.50 m) New South Wales Certain ARTC lines after about 2012.
- 4.20 m (13 ft 9 in) LGV Rhône-Alpes.
- 4.23 m (13 ft 11 in) Tōkaidō Shinkansen.
- 4.30 m (14 ft 1 in) San›yō Shinkansen.
- 4.50 m (14 ft 9 in) Erfurt–Leipzig/Halle high-speed railway.
- 4.70 m (15 ft 5 in) Nuremberg–Erfurt high-speed railway.

## Measurement

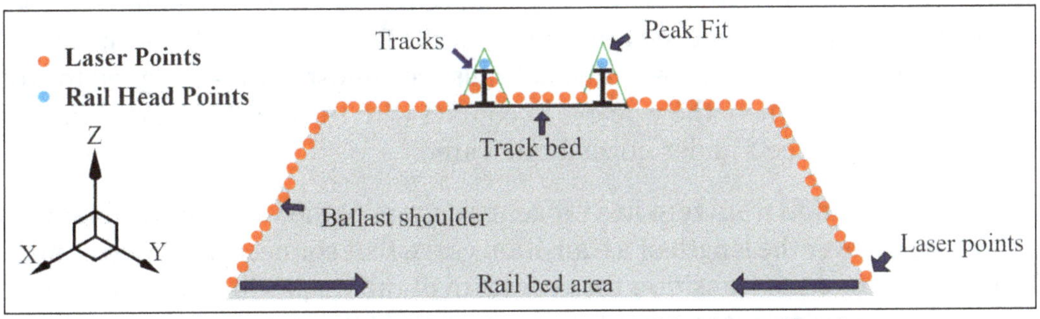


Measurement of Track Spacing from the rail head to rail head.

By definition the track spacing is given from center to center of a rail track. For an actual construction the distance is measured from the inside of a rail head to the matching one of the other track. As far as both tracks have the same gauge this is the same distance.

## Track Transition Curve

A track transition curve, or spiral easement, is a mathematically-calculated curve on a section of highway, or railroad track, in which a straight section changes into a curve. It is designed to prevent sudden changes in lateral (or centripetal) acceleration. In plane (viewed from above), the start of the transition of the horizontal curve is at infinite radius, and at the end of the transition, it has the same radius as the curve itself and so forms a very broad spiral. At the same time, in the vertical plane, the outside of the curve is gradually raised until the correct degree of bank is reached.

If such an easement were not applied, the lateral acceleration of a rail vehicle would change abruptly at one point (the tangent point where the straight track meets the curve), with undesirable results. With a road vehicle, the driver naturally applies the steering alteration in a gradual manner, and the curve is designed to permit that by using the same principle.

## Geometry

While railroad track geometry is intrinsically three-dimensional, for practical purposes the vertical and horizontal components of track geometry are usually treated separately.

The overall design pattern for the vertical geometry is typically a sequence of constant grade segments connected by vertical transition curves in which the local grade varies linearly with distance and in which the elevation therefore varies quadratically with distance. Here grade refers to the tangent of the angle of rise of the track. The design pattern for horizontal geometry is typically a sequence of straight line (i.e., a tangent) and curve (i.e. a circular arc) segments connected by transition curves.

The degree of banking in railroad track is typically expressed as the difference in elevation of the two rails, commonly quantified and referred to as the superelevation. Such difference in the elevation of the rails is intended to compensate for the centripetal acceleration needed for an object to move along a curved path, so that the lateral acceleration experienced by passengers/the cargo load will be minimized, which enhances passenger comfort/reduces the chance of load shifting (movement of cargo during transit, causing accidents and damage).

It is important to note that superelevation is not the same as the roll angle of the rail which is used to describe the “tilting” of the individual rails instead of the banking of the entire track structure as reflected by the elevation difference at the “top of rail”. Regardless of the horizontal alignment and the superelevation of the track, the individual rails are almost always designed to “roll”/”cant” towards gage side (the side where the wheel is in contact with the rail) to compensate for the horizontal forces exerted by wheels under normal rail traffic.

The change of superelevation from zero in a tangent segment to the value selected for the body of a following curve occurs over the length of a transition curve that connects the tangent and the curve proper. Over the length of the transition the curvature of the track will also vary from zero at the end abutting the tangent segment to the value of curvature of the curve body, which is numerically equal to one over the radius of the curve body.

The simplest and most commonly used form of transition curve is that in which the superelevation and horizontal curvature both vary linearly with distance along the track. Cartesian coordinates of points along this spiral are given by the Fresnel integrals. The resulting shape matches a portion of an Euler spiral, which is also commonly referred to as a “clothoid”, and sometimes “Cornu spiral”.

A transition curve can connect a track segment of constant non-zero curvature to another segment with constant curvature that is zero or non-zero of either sign. Successive curves in the same direction are sometimes called progressive curves and successive curves in opposite directions are called reverse curves.

The Euler spiral provides the shortest transition subject to a given limit on the rate of change of the track superelevation (i.e. the twist of the track). However, as has been recognized for a long time, it has undesirable dynamic characteristics due to the large (conceptually infinite) roll acceleration and rate of change of centripetal acceleration at each end. Because of the capabilities of personal computers it is now practical to employ spirals that have dynamics better than those of the Euler spiral.

## Hallade Method

The Hallade method, devised by Frenchman Emile Hallade, is a method used in track geometry for surveying, designing and setting out curves in railway track.

It involves measuring the offset of a string line from the outside of a curve at the central point of a chord. A standard chord length is used: in the UK this is conventionally one chain, i.e. 22 yards. Half chords, i.e. half a chain intervals, are marked on the datum rail using chalk. The string, which is one chain long, is then held taut with one end on two marks at each end of a chord, and the offset at the half chord mark measured. The versine of the chord, which is equal to this measured offset value can be calculated using the approximation of:

$$\text{versin}\,\theta = 1 - cos\theta$$

which is:

$$v \approx \frac{L^2}{8r}$$

where v = versine (m), L = chord length (m), r = radius of curve (m).

This formula is also true for other units of measurement such as in feet. The relationship of versine, chord and radius is derived from the Pythagorean theorem. Based on the diagram on the right:

$$OC = \sqrt{OA^2 - AC^2}$$

We can replace OC with r (radius) minus v, OA with r and AC with L/2 (half a chord). Then the rearrange formula to:

$$r - v = \sqrt{r^2 - (\frac{L}{2})^2},$$

$$r^2 - 2rv + v^2 = r^2 - (\frac{L}{2})^2,$$

$$2r - v = \frac{L^2}{4v},$$

$$r = \frac{L^2}{8v} + \frac{v}{2}.$$

Since the curved tracks are usually large, the result of v/2 is very small. To simplify the formula, the approximation is:

$$r \approx \frac{L^2}{8v}$$

The following can be used to find the versine of a given constant radius curve:

$$v \approx \frac{L^2}{8r}$$

The Hallade method is to use the chord to continuously measure the versine in an overlapping pattern along the curve. The versine values for the perfect circular curve would have the same number. By comparing the surveyed versine figures to the design versines, this can then be used to determine what slues should be applied to the track in order to make the curve correctly aligned. This is often done using pegs which are driven into the ground in the cess beside the track to be aligned. The process of putting the pegs in the correct positions is known as 'setting out'. If the curve needs to be of a desired constant radius, which will usually be determined by physical obstructions and the degree of cant which is permitted, the versine can be calculated for the desired radius using this approximation. In practice, many track curves are transition curves and so have changing radii. In order to maintain a smooth transition, the differences in versines between consecutive half chords are measured and minimised.

The Hallade survey is a survey method that uses the same principle to measure the versines along an existing curve. Based on the versine values, the radius of that circular curved track can be approximated to:

$$r \approx \frac{L^2}{8v}$$

This method can be done manually, and this method is still used in the UK. However, due to the complexity of the calculations over long lengths of track, it is now often done by computer, with the track geometry data being loaded straight onto a computer controlled tamping and lining machine for implementation.

## Track Ballast

Track ballast forms the trackbed upon which railroad ties (sleepers) are laid. It is packed between, below, and around the ties. It is used to bear the load from the railroad ties, to facilitate drainage of water, and also to keep down vegetation that might interfere with the track structure. This also serves to hold the track in place as the trains roll by. It is typically made of crushed stone, although ballast has sometimes consisted of other, less suitable materials, for example burnt clay. The term "ballast" comes from a nautical term for the stones used to stabilize a ship.

### Construction

The appropriate thickness of a layer of track ballast depends on the size and spacing of the ties, the amount of traffic on the line, and various other factors. Track ballast should never be laid down less than 150 mm (6 inches) thick; and high-speed railway lines may require ballast up to ½ metre (20 inches) thick. An insufficient depth of ballast causes overloading of the underlying soil, and in unfavourable conditions overloading the soil causes the track to sink, usually unevenly. Ballast less than 300 mm (12 inches) thick can lead to vibrations that damage nearby structures. However, increasing the depth beyond 300 mm (12 inches) adds no extra benefit in reducing vibration.

In turn, track ballast typically rests on a layer of small crushed stones: the sub-ballast. The sub-ballast layer gives a solid support for the top ballast, and reduces the seepage of water from the underlying ground. Sometimes an elastic mat is placed on the layer of sub-ballast and beneath the ballast, thereby significantly reducing vibration.

It is essential for ballast to be piled as high as the ties, and for a substantial "shoulder" to be placed at their ends; the latter being especially important, since this ballast shoulder is the main restraint of lateral movement of the track. The ballast shoulder always should be at least 150 mm (6 inches) wide, and may be as wide as 450 mm (18 inches).

Ballast must be irregularly shaped to work properly.

The shape of the ballast is also important. Stones must be irregularly cut, with sharp edges, so that they properly interlock and grip the ties in order to fully secure them against movement; spherical stones cannot do this. In order to let the stones fully settle and interlock, speed limits are often lowered on sections of track for a period of time after new ballast has been laid.

## Maintenance

New track ballast ready for laying at Boxmeer railway station.

A ballast regulator shaping newly placed ballast.

Ballast tamping machine as used in railroad track maintenance.

If ballast is badly fouled, the clogging will reduce its ability to drain properly; this, in turn, causes more debris to be sucked up from the sub-ballast, causing more fouling. Therefore, keeping the ballast clean is essential. Bioremediation can be used to clean ballast.

It is not always necessary to replace the ballast if it is fouled, nor must all the ballast be removed if it is to be cleaned. Removing and cleaning the ballast from the shoulder is often sufficient, if shoulder ballast is removed to the correct depth. While this job was historically done by manual labour, this process is now, like many other railway maintenance tasks, a mechanised one, with a chain of specially-designed railroad cars handling the task. One wagon cuts the ballast and passes it via a conveyor belt to a cleaning machine, then the cleaning wagon washes the ballast, and deposits the dirt and ballast into other wagons for disposal and re-use, respectively. Such machines can clean up to two kilometres (1.2 mi) of ballast in an hour.

Cleaning, however, can only be done a certain number of times before the ballast is damaged to the point that it cannot be re-used; furthermore, track ballast that is completely fouled can not be corrected by shoulder cleaning. In such cases, it is necessary to replace the ballast altogether. One method of "replacing" ballast, if necessity demands, is to simply dump fresh ballast on the track, jack the whole track on top of it, and then tamp it down; alternatively, the ballast underneath the track can be removed with an undercutter, which does not require removing or lifting the track.

The dump and jack method cannot be used through tunnels, under overbridges, and where there are platforms. Where the track is laid over a swamp, such as the Hexham swamp in Australia, the ballast continuously sinks, and needs to be topped up to maintain its line and level. After 150 years of topping up, there appears to be 10 m (33 ft) of sunken ballast under the tracks. Chat Moss in the United Kingdom is similar.

Regular inspection of the ballast shoulder is important; as noted earlier, the lateral stability of the track depends upon the shoulder. The shoulder acquires some amount of stability over time, being compacted by traffic; maintenance tasks such as replacing ties, tamping, and ballast cleaning can upset this stability. After performing these tasks, it is necessary for either trains to run at reduced speed on the repaired routes, or to employ machinery to compact the shoulder again.

If the trackbed becomes uneven, it is necessary to pack ballast underneath sunk ties to level the track out again. This is, in the mechanized age, usually done by a ballast tamping machine. A more recent, and probably better, technique is to lift the rails and ties, and to force stones, smaller than the track ballast particles and all of the same size, into the gap. This has the advantage of not disturbing the well-compacted ballast on the trackbed, as tamping is likely to do. This technique is called pneumatic ballast injection (PBI), or, less formally, "stoneblowing". However, this technique is not as effective with fresh ballast, as the smaller stones tend to move down between the larger pieces of ballast.

## Quantities

The quantity of ballast tends to vary with gauge, with the wider gauges tending to have wider formations. The depth of ballast also tends to vary with the density of traffic, as faster and heavier traffic requires greater stability. The quantity of ballast also tends to increase over the years as more and more ballast is piled on. Some figures from an 1897 report are:

- First class line – 60 lb/yd (29.8 kg/m) rail – 1,700 cu yd/mi (810 m³/km).
- Second class line – 41.5 lb/yd (20.6 kg/m) rail – 1,135 cu yd/mi (539 m³/km).
- Third class line – 30 lb/yd (14.9 kg/m) rail – 600 cu yd/mi (290 m³/km).

## Tamping Machine

A tamping machine or ballast tamper is a machine used to pack (or tamp) the track ballast under railway tracks to make the tracks more durable. Prior to the introduction of mechanical tampers, this task was done by manual labour with the help of beaters. As well as being faster, more accurate, more efficient and less labour-intensive, tamping machines are essential for the use of concrete sleepers since they are too heavy (usually over 250 kg (551 lb)) to be lifted by hand.

Early machines only lifted the track and packed the ballast. More modern machines, sometimes known as a tamper-liner or tamping and lining machine, also correct the alignment of the rails to make them parallel and level, in order to achieve a more comfortable ride for passengers and freight and to reduce the mechanical strain applied to the rails by passing trains. This is done by finding places where the sleepers have sunk from the weight of the passing trains or frost action, causing the track to sag. The tamper lifts each sleeper and the rails up, and packs ballast underneath. When the sleeper is laid down again, the sagged rails now sit at the proper level. In cases where frost action has caused adjacent rails to rise higher, ballast tampers can raise rails above their original level to make the line level again. "Lining" rails doesn't involve ballast tamping, it merely ensures the rails are perfectly parallel and straight as possible. Combining tamping and lining into a single machine saves time and money, as only one machine needs to be run over the track to perform both functions.

## Functions

A Balfour Beatty ballast tamper/track tamper train at Banbury station.

- Packing of ballast under sleepers,
- Correction of cross level,
- Correction of longitudinal level,
- Driving,
    - Run drive,
    - Work drive,
- Special functions,
    - Twist corrections,
    - Track settlement,
    - Laser lining,
    - Automatic data feeding by computer,
    - Recording of track parameter.

## Types

A GTRM Rail tamping machine.

An Amey Plc tamping machine.

Tamper machines are built in many different varieties depending on their purpose:

- Plain line tamping machines: Used on line sections which have no points or other complex track structures, commonly referred to as production machines, generally have fixed tamping head positions:
    - Duomatic
    - Continuous action tamping machines

- Tamping Express

- Switch tamping machines: Used to tamp switches, crossings and other complex track structures, have movable tamping heads with ability to isolate heads when required.

## Tamping Process

The tamping process from any type of tamper consists of the following basic steps:

- Lifting, lining unit moves rail and sleepers under tamping unit into correct vertical and horizontal position.
- Tamping units drive vibrating tines into the ballast on both sides of the sleeper until squeezing depth is achieved.
- Vibrating tines are pushed together to pack ballast under lifted, lined sleeper to ensure it holds position when released by the lifting, lining unit.
- Tamping unit retracts to stowage position, track is released by lifting, lining unit.
- Machine moves to the next sleeper and begins a new cycle.

## Standard Tamping Machine Arrangement

The basic principles and functions of a tamping machine remain the same regardless of manufacturer with only minor differences in design.

## Drive

The majority of track machines are powered by a diesel engine. This provides power to the driving wheels via either a hydrostatic circuit or cardan shaft, allowing the machine to propel itself to and around a work-site. The engine also drives a hydraulic pump to provide power for the various tools.

## Lifting Lining Unit

The lifting lining unit of a tamping machine is used to lift and hold the track in corrected position while being tamped. All types of units require the following components to achieve this task:

- Chassis, normally rectangular with 4 wheels to ride on the rails.
- Lifting cylinders, 1 cylinder on either side connecting the unit to main machine chassis and allowing unit to lift track vertically.
- Lining cylinders, 2 horizontal cylinders between unit chassis and main machine chassis used to push/pull the track horizontally.
- Clamp or "hook" assemblies, these are usually on the outside of each rail and utilize hydraulic cylinders to clamp onto the side of the rail head so that is locked to the unit chassis before lifting and lining is commenced.

The lifting lining unit is usually secured to the main machine chassis via an hydraulically length adjustable trailing arm, The arm is adjustable so that the unit can be moved clear of small obstructions such as insulated joints or wires.

## Reference System

To ensure track geometry is corrected, most tamping machines use a two chord lining system (one for vertical alignment and one for horizontal alignment). The two chord system requires three reference trolleys fitted to the machine - usually called A point, B point and C point. Some machines will use a fourth point between B and C to carry out quality control measurements.

A point is always the front reference point and sits on uncorrected track. The anchorage for chords on A point are able to move to compensate for defects in track geometry - this is done by the "tower" operator. Depending on the chord system, the anchorages will be either a wire anchor or a light source.

B point is positioned as close as physically possible to the lifting lining unit. B point is used by the machine control system to position the track correctly using either potentiometers or optical filters, depending on type of chord system used.

C point is the rear-most measuring point and anchorage point for lifting and lining chords. Depending on type of chord system C point will be either wire anchor with a tensioning cylinder or a photoelectric light receiver.

All three points are individual rail trolleys able to freely move up, down, left and right independent of the machine chassis and hence follow any minor fluctuation in rail position. When working, the machine uses pneumatic cylinders to lightly push these trolleys into the selected datum rail both vertically and horizontally.

When using this method, the tower operator positions A point anchorages according to existing track geometry measurements taken beforehand. Once A point has been positioned, it is then assumed that both A and C are in correct position as if machine were sitting on corrected track. The machine then uses the lifting lining unit to move the rail and B point in line with A and C.

## Tamping Units

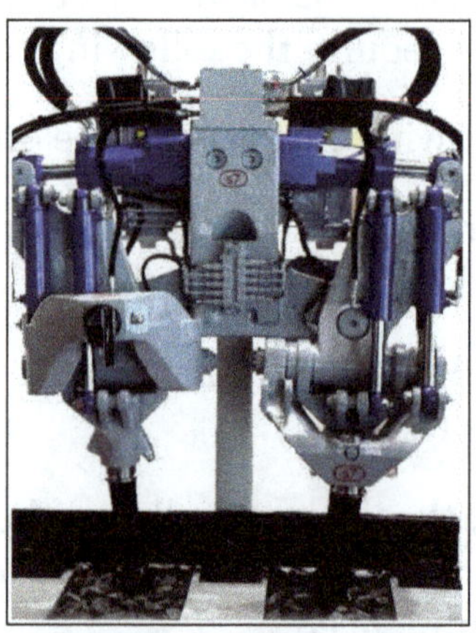

System 7 fully hydraulic tamping head.

The Tamping units of most Tamping machines will consist of:

- A set of tines either 8 or 16 per sleeper divided evenly between front and rear of sleeper.
- Mounting or "guide" rods that allow the unit as a whole to move up and down in a linear motion.
- Support frame.
- Squeezing arms.
- Tine mounts, can be either part of the squeezing arm or attached to it via a pin to allow swiveling depending on type of tamping machine.

To generate the vibration needed for penetration and consolidation there are two leading methods commonly used:

- Using a hydraulically driven eccentric shaft attached either to the end of the squeezing cylinders or acting as the pivot point for the squeezing arms.
- Using a special squeezing cylinder and valve assembly to oscillate the tines.

A less common method more often seen on tamping head attachments for excavators is to use a motor driven vibrator assembly that is directly bolted to the support frame.

## Specialist Tamping Machines

## Continuous Action Tamping Machines

A *Continuous Action Tamping Machine* (CAT) can pack between one and four sleepers at a time, with outputs between 320m/h and 2600m/h generally anticipated.

## Dynamic Tamping Express

The "Tamping Express" is a machine developed by Plasser & Theurer, and in the UK & Europe is referred to as the 09-3X. This machine consists of a conventional CAT styllish satellite with tooling for 3 sleepers in continuous succession, along with a full DTS stabilising unit suspended from the rear most vehicle in the machine.

## DGS/DTS (Dynamic Track Stabilising)

## Tamper in Jarvis PLC Fastline Livery

Tamping and cleaning operations have the adverse effect of reducing the resistance of the track to lateral movement. The resistance gradually recovers with the passage of trains, but may require a speed restriction imposed for the duration. This 'consolidation' can be achieved faster and in a more controlled manner using mechanised equipment known as the Dynamic Track Stabiliser (DTS).

A DTS will normally be used only after a stretch of track has been tamped and aligned.

D.G.S. has a vibrating unit which holds the track in position and applies a horizontal vibration and vertical load to simulate the passage of trains. The track parameters (or *cross levels*), before and after stabilising, can be viewed through bogies in the front and rear.

Dynamic Track Stabilising has the following advantages, resulting in enhanced safety:

- Increases the resistance of the track to lateral movement.
- Creates a homogenous bed of ballast.
- Permits earlier relaxation of speed restrictions.
- Eliminates non-uniform initial differential settlements caused by rail traffic.
- Retains correct track geometry for a longer period than was achievable using tamping machines alone.
- Vibration frequency: 30–35 Hz.
- Working speed: 1–2 km/h.
- Vertical pressure applied: 100 kg/cm$^2$.

The stabilisation achieved by one pass of a D.G.S. is equal to that achieved by 100,000 tonnes of traffic, and allows a speed restriction of 20 km/h to be relaxed to 40 km/h.

Dynamic stabilisation is usually avoided on bridges or around overhead structures since there is a risk of damage to foundations.

## Design of Ballast Section

## Minimum Depth of Ballast Cushion

The load on the sleeper is transferred through the medium of the ballast to the formation. The pressure distribution in the ballast section depends upon the size and shape of the ballast and the degree of consolidation. Though the lines of equal pressure are in the shape of a bulb, yet for simplicity, the dispersion of load can be assumed to be roughly 45° to the vertical. In order to ensure that the load is transferred evenly on the formation, the depth of the ballast should be such that the dispersion lines do not overlap each other.

For the even distribution of load on the formation, the depth of the ballast is determined by the following formula.

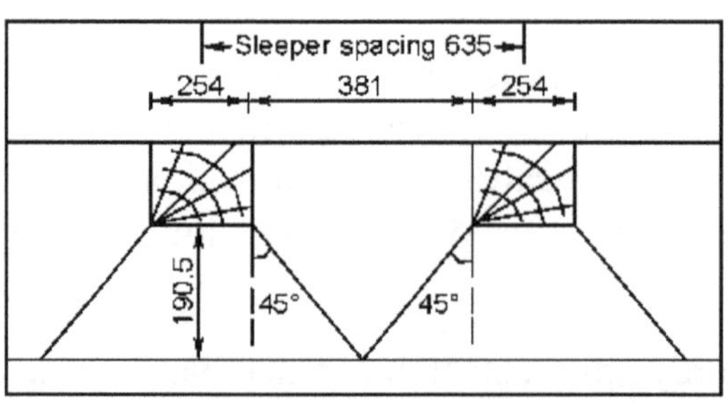


Minimum depth of ballast cushion (dimensions in mm).

Sleeper spacing = width of the sleeper + 2 x depth of ballast (8.1) If a BG track is laid with wooden sleepers with a sleeper density of N + 6, then the sleeper spacing would be 68.4 cm. If the width of the sleeper is 25.4 cm, then the depth of the ballast cushion would be

$$d=\frac{68.4-25.4}{2}=21.5\text{cm}$$

A minimum cushion of 15-20 cm of ballast below the sleeper bed is normally prescribed on Indian Railways.

## Ballast Profile for Fish-plated Track

The ballast profile for a fish-plated track is shown in figure. The requirements of ballast for different groups of railway lines as adopted by Indian Railways are given in table.

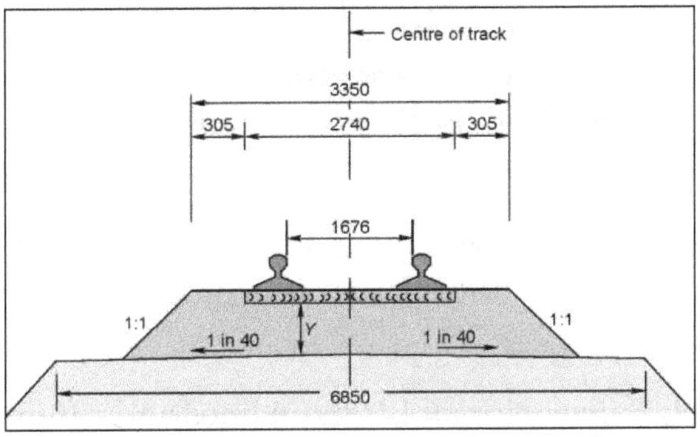


Ballast requirement for fish-plated tracks.

| Group | Recommended depth of ballast cushion Y (mm) | Quantity of ballast required/metre | |
|---|---|---|---|
| | | On straight lines and curves of radius greater than 600 m ($m^3$) | On curves of radius lower than 600 m ($m^3$) |
| A | 300 | 1.588 | 1.634 |
| B and C | 250 | 1.375 | 1.416 |
| D | 200 | 1.167 | 1.202 |
| E | 150 | 0.964 | 0.996 |

In the case of ordinary fish-plated tracks, to be increased on the outside of the curves to 400 mm in the case of sharper curves of a radius more than 600 m. In short welded panel tracks, it is to be increased to 400 mm on the outside of all curves flatter than 875 m and to 450 mm in the case of sharper curves with a radius more than 875 m. To be increased to 550 m on the outside of the turn on curves of turnouts in passenger yards. In the case of a short welded rail (SWR) track, the minimum depth of cushion should be 200 mm.

The ballast profile for a long welded rail (LWR) track is shown. The minimum clean stone ballast cushion below the bottom of sleeper (A) is 250 mm. For routes where speeds are to be more than 130 kmph, A is 300 mm-200 mm along with 150 mm of sub-ballast. Suitable dwarf walls should be provided in the case of cuttings, if necessary, for retaining the ballast.

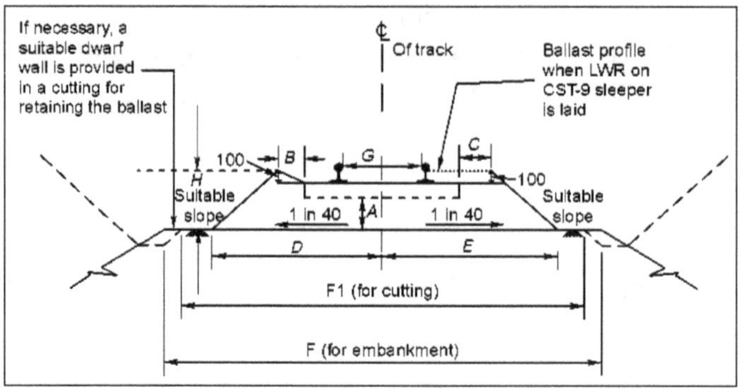


Ballast requirement (mm) for single-line BG LWR tracks.

## Specifications for Track Ballast

The ballast material should possess the general qualities described below:

- Basic quality: The ballast should be hard, durable, as far as possible angular along edges/ corners, and free from weathered portions of parent rock, organic impurities, and inorganic residues.
- Particle shape: Ballasts should be cubical in shape as far as possible. Individual pieces should not be flaky and should have flat faces generally with not more than two rounded/ sub-rounded faces.
- Mode of manufacture: Ballasts for all BG main lines and running lines, except on E routes, but including E special routes, should be machine crushed. For other BG lines and MG/ NG routes planned or sanctioned for conversion, the ballast should preferably be machine crushed. Hand broken ballast can be used in exceptional cases with the prior approval of the chief track engineer or the CAO (chief administrative officer). Such approval should be obtained prior to the invitation of tenders. Hand broken ballasts can be used without any formal approval on MG and NG routes not planned or sanctioned for conversion.

## Physical Properties

All ballast samples should possess the physical properties given in when tested in accordance with IS:2386 (IV)-1963.

Table: Physical Requirements of Ballast.

| Characteristics | BG, MG, and NG (planned/sanctioned for conversion) | NG and MG (other than those planned for conversion) |
|---|---|---|
| Aggregate abrasion | 30% maximum | 35% maximum |
| Aggregate impact | 20% maximum | 30% maximum |
| Water absorption | 1% | - |

In exceptional cases, relaxable on technical and/or economic grounds up to 35% and 25%, respectively, by the chief track engineer (CTE) in open lines and the chief administration officer (construction) (CAO/C) for construction projects. Relaxation in abrasion and impact values is given prior to the invitation of tender and should be incorporated in the tender document.

## Size and Gradation

The ballast should satisfy the size and gradation requirements given in table.

Table: Ballast gradation.

| Size of sieve | % retained |
|---|---|
| 65 mm | 5% maximum |
| 40 mm | 40% to 60% |
| 20 mm | Not less than 98% for machine crushed and not less than 95% for hand broken |

## Oversized Ballast

- Retention on 65-mm square mesh sieve A maximum of 5% ballast retained on a 65-mm sieve is allowed without deduction in payment. In case the ballast retained on a 65-mm sieve exceeds 5% but is less than 10%, payment at a 5% reduction of 5% in the contracted rate is made for the full stack. Stacks retaining more than 10% of ballast on a 65-mm sieve are rejected.
- Retention on 40-mm square mesh sieve In case the ballast retained on a 40-mm square mesh sieve (machine crushed only) exceeds the 60% limit prescribed above, payment at the following reduced rates is made for the full stack in addition to the reduction as worked out above:
    - 5% reduction in contracted rates if the retention on a 40-mm square mesh sieve is between 60% (excluding) and 65% (including).
    - 10% reduction in contracted rates if retention on a 40-mm square mesh sieve is between 65% (excluding) and 70% (including).
    - In case the retention on a 40-mm square mesh sieve exceeds 70%, the stack is rejected.
    - In the case of hand broken ballast supply, 40-mm-sieve analysis may not be carried out. The executive may, however, ensure that the ballast is well graded between 65 mm and 20 mm.

## Undersized Ballast

The ballast is treated as undersized and rejected if:

- Retention on a 40-mm sieve is less than 40%.
- Retention on a 20-mm sieve is less than 98% (for machine crushed ballast) or 95% (for hand broken ballast).

## Shrinkage Allowance

Payment is made for the gross measurements either in stacks or in wagons without any deduction for shrinkage/voids. However, when ballast is supplied in wagons, up to 8% shrinkage is permitted at the destination by the consignee verifying the booked quantities.

## Sampling and Testing

The following procedure is specified for the sampling and testing of the ballast:

a) A minimum of three samples of ballast should be taken for sieve analysis for measurement done on any particular date, even if the number of stacks to be measured is less than three.

b) The tests for abrasion value, impact value, and water absorption should be done in approved laboratories or in the Railways' own laboratories.

c) In order to ensure the supply of a uniform quality of ballast, the specifications given in Table 8.6 should be followed with respect to sampling, testing, and acceptance of the ballast. The tests given in this table may be carried out more frequently if warranted at the discretion of the Railways.

d) On supply of the first 100 m³, tests for size gradation, abrasion value, impact value, and water absorption (if prescribed) should be carried out by the Railways. Further supply should be accepted only after the first batch satisfies these tests. The Railways reserves the right to terminate the contract as per the general conditions of contract (GCC) at this stage itself in case the ballast supply fails to meet any of these specifications.

Table: Frequency of tests for ballast supply.

| Item | Supply in | Stacks | Supply in wagons |
|---|---|---|---|
| | For a stack of | For a stack of | |
| | volumne less | volume more | |
| | than 100 $m^3$ | than 100 $m^3$ | |
| Number of size | One for each stack | One for each stack | One for each wagon |
| and gradation tests | | | |
| Size of one | 0.027 | 0.027 for every | 0.027 |
| sample ($m^3$) | | 100 $m^3$ or part | |
| | | thereof | |
| Abrasion value, impact | | | |
| Value, and water | One test for every 2000 $m^3$ of ballast | | |
| Absorption test | | | |

This sample should be collected using a wooden box of internal dimensions 0.3 m x 0.3 m x 0.3 m from different parts of the stack/wagon.

These tests should be done for the purpose of monitoring the quality during supply. In case the test results are not as per the prescribed specifications at any stage, further supplies should be suspended till suitable corrective action is taken and supply as per specifications is ensured.

## Functions of Ballast

Some of the important functions of railway ballast are:

1. To provide firm and level bed for the sleepers to rest on.
2. To allow for maintaining correct track level without disturbing the rail road bed.
3. To drain off the water quickly and to keep the sleepers in dry conditions.
4. To discourage the growth of vegetation.
5. To protect the surface of formation and to form an elastic bed.
6. To hold the sleepers in position during the passage of trains.
7. To transmit and distribute the loads from the sleepers to the formation.
8. To provide lateral stability to the track as a whole.

## Requirements for Ideal Ballast

The ideal material for ballast should fulfill the following requirements:

1. It should be possible to maintain the required depth of the material in order to distribute the load of passing train on the formation ground.
2. The material to be used for ballast should not be too rigid but it should be elastic in nature.
3. The material for ballast should be of such nature that it grips the sleepers in position and prevent their horizontal movement during passage of train.
4. It should not allow the rain water to accumulate but should be able to drain off the water immediately without percolating.
5. It should be strong enough a resistance to abrasion.

## Materials for Ballast

The following materials are used for ballast on the railway track:

1. Broken Stone.
2. Gravel.
3. Cinders/Ashes.
4. Sand.
5. Kankar.
6. Moorum.
7. Brick Ballast.
8. Selected Earth.

# Railroad Tie

Railway sleepers, also called railroad ties, railway ties or crossties, are an important railway component. Generally the rail sleeper is always laying between the two rail tracks to keep the correct space of gauge. What are railway sleepers made of? Having been developed for more than one hundred years, the railway ties need to meet the different requirements of the various railway tracks. In the past time, railroad ties were usually made of wood and had continued for about 50 years. Then with the development of steel tracks, steel sleepers appeared. And later, the first concrete sleeper experiment was made in Germany in 1906 between the line Nuremberg and Bamberg. In the recent time, concrete sleepers are widely used, especially in Europe and Asia. And in UK, steel ties are common. Besides, plastic composite ties are also employed in the rail track transportation.

## Types of Sleeper

### Wooden Sleeper

Historically wooden ties were made of a variety of softwood and some popular hardwoods such as oak, jarrah, karri etc. And they are only suitable for low speed lines with the limit speed of 160 km/h. As to acceptable species of wood for sleepers, the types are European oak, beech, pine etc. But nowadays, wooden sleepers are mostly replaced by concrete sleepers in some countries.

### Advantage

- Easy to manufacture and handle Well electrical insulation.
- Easily adapted to non-standard situations.

### Disadvantages

- Non reusable.
- Expensive of the limit wood resource.

### Steel Sleeper

In recent times, steel sleepers mostly handle the heavy loads and can be designed to suit different rail track specifications. Because of the same weight as timber, steel sleeper can replace wood sleepers and be used on a ballasted bridge, providing a more durable and stronger solution without increasing bridge load.

## Advantages

- Easy to install and manufacture.
- Handle more weight.

## Disadvantages

- Sensitive to the chemical attacks.
- Hard to maintenance.
- Low transverse resistance.

## Concrete Sleeper

Compared with wooden sleepers, which are increasingly difficult and expensive to source in sufficient quantities and quality, concrete ties are cheaper and easier to obtain. In general, it can also be divided into pre-stressed mono block concrete sleeper and reinforced twin block concrete sleeper. Due to the greater weight which helps sleepers remain in the correct position longer, concrete ties require less maintenance than timber and own a longer service life. In some countries, concrete sleepers occupy a very important position. For example, on the highest categories of line in the UK, pre-stressed concrete ties are the only one permitted by Network Rail standards.

## Advantages

- Cheaper.
- Easier to obtain.
- Less maintenance.
- Longer service life.

## Disadvantages

- Hard to handle due to large weight.
- Difficult to maintain longitudinal level due to higher.
- Inertia moment and lower elasticity.

## Sleeper Density and Spacing of Sleepers

Sleeper density is the number of sleepers per rail length. It is specified as M+x or N+x, where M or N is the length of the rail in metres and x is a number that varies according to factors such as:

- Axle load and speed.
- Type and section of rails.
- Type and strength of the sleepers.
- Type of ballast and ballast cushion.
- Nature of formation.

If the sleeper density is M + 7 on a broad gauge route and the length of the rail is 13 m, it means that 13 + 7 = 20 sleepers will be used per rail on that route. The number of sleepers in a track can also be specified by indicating the number of sleepers per kilometre of the track. For example, 1540 sleepers/km. This specification becomes more relevant particularly in cases where rails are welded and the length of the rail does not have much bearing on the number of sleepers required. This system of specifying the number of sleepers per kilometre exists in many foreign countries and is now being adopted by Indian Railways as well.

The spacing of sleepers is fixed depending upon the sleeper density. Spacing is not kept uniform throughout the rail length. It is closer near the joints because of the weakness of the joints and impact of moving loads on them. There is, however, a limitation to the close spacing of the sleepers, as enough space is required for working the beaters that are used to pack the joint sleepers. The standard spacing specifications adopted for a fish-plated track on Indian Railways are given in table.

Table: Spacing of sleepers for a fish-plated track.

| Spacing of sleepers | Broad gauge | Meter gauge | | |
|---|---|---|---|---|
| | centre-to-centre | centre-to-centre | | |
| | spacing (mm) | spacing (mm) | | |
| | Wooden | Metal | Wooden | Metal |
| Between joint sleepers (a) | 300 | 380 | 250 | 330 |

| Between joint sleepers and the first | 610 | 610 | 580 | 580 |
|---|---|---|---|---|
| shoulder sleeper (b)<br>Between first shoulder sleeper and | 700 | 720 | 700 | 710 |
| second shoulder sleeper (c) for sleeper density M + 4 | (640) | (630) | (620) | (600) |
| Between intermediate sleepers (d) | 840 | 830 | 820 | 810 |
| for sleeper density M + 4 | (680) | (680) | (720) | (640) |

Values within parentheses are those for sleeper density M + 7.

Table: Spacing of sleepers for welded track.

| No. of sleepers per km | Exact centre-to-centre spacing required as per calculation (mm) | Centre-to-cen be provided in LWR track | itre spacing to the field (mm) |
|---|---|---|---|
| 1660 | 602.4 | 600 | SWR track |
| 1540 | 649.3 | 650 | 660 |
| 1310 | 763.3 | - | 780 |

Now-a-days sleeper density is also indicated in terms of the number of sleepers/km. The sleeper spacing required for various sleeper densities is given in figure.

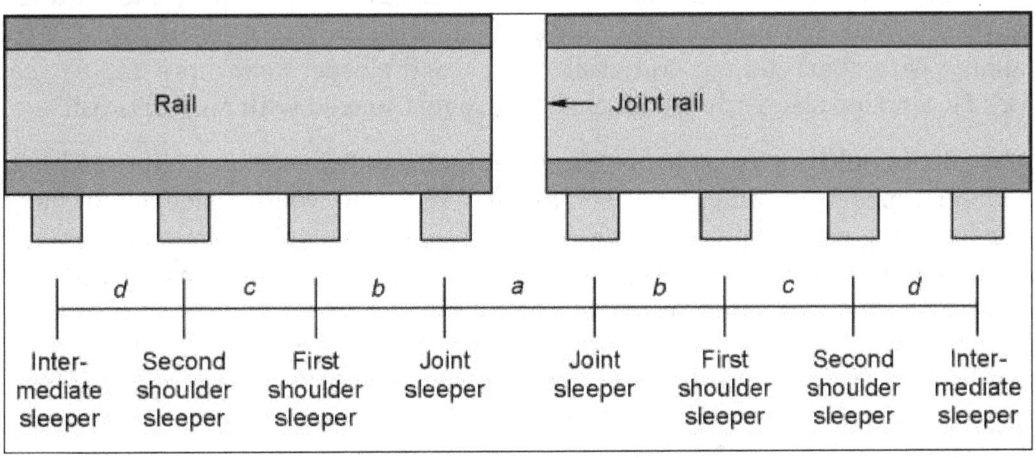


Spacing of sleepers on a fish-plated track.

## Advantages of Wooden Sleeper

- Good elasticity: The stiffness (elasticity) under the rail is an important indicator of the track structure. Poor flexibility will cause extra damage to the rails and vehicles. The elasticity of the wooden sleeper is very good, and it is not even necessary to use the elastic spacer under the rail directly.
- Simple processing, transportation and laying: The wood can be directly processed into the

required size and then the antiseptic solution can be soaked. It is also convenient because it is lighter to transport. Laying is the same reason, light and easy to transport the nature will be laid.

- Easy to connect with rails: The previous wooden sleepers were all made directly by a few dog spikes. Even if they were fixed, it was much simpler than the current type of fasteners. Of course, it is also a disadvantage, and it is not very reliable.
- Good insulation: The combination of wood + anti-corrosion oil, but how good the insulation in the end I.

### Disadvantages of Wood Sleeper

- Consumption of wood resource is not environmentally friendly. This needless to say, the basic wooden sleeper is grown up in the mountains, although it is now possible to use rotation and other methods to reduce the impact on the environment, but after all, there is still impact? It's okay to take the wood to make furniture.
- Uniformity is not good, uneven elasticity and strength. This is quite understandable. Different trees have different elasticities. Different locations of the same tree are also different. It is like that there are no two identical leaves, and there are no two sleepers. After all, natural growth is not the same model. from. This brings about the uneven stiffness under the rail, which is very unfavorable for the relatively fast running line, so now the wooden sleeper is mainly used in the depot, and the main line is basically unused.
- Erodible, short service life. Even it is soaked in preservative oil, and to be thoroughly soaked (now requiring a immersion depth of more than 13mm, different grade requirements are slightly different, not meticulous), but after all it is still wood, hardness, strength are not available with concrete, etc. Material ratio, wood sleepers can now use 10 years is very good. Ps, wooden sleeper looks black, because it is soaked with antiseptic oil.
- The ability to hold dog spike is poor, and geometry is difficult to maintain. The form of the wooden sleeper fixed rail can be seen as the nail to the wooden sleeper (in fact, there are other fastening parts), the wheel will have a lateral force on the spike, the spike will move so are the rails, so that the gauge, which can be simply understood as the distance between the two rails, changes, and you need to adjust.

## Rail Fastening System

A rail fastening system is used to fix rails to railway sleeper or railroad ties, which is usually consisted of rail anchors, rail tie plates, chairs, fasteners, spikes, screws, track bolts, etc. The selection of a fastening system must be made with utmost care for rail road safety can't be compromised.

### Types of Fastenings

Various types of fastenings have been developed to meet the requirements of railway development. Namely, there are fastening with rail spike with base plate above the tie, E-clip Fastening, Pandrol

Fast Clip Fastening, Tension Clamp Fastening, Bolt Clamp Fastening, Tack Joint and Chairs Fastening, and Steel Spring Keyed rail in Chair Fastening, the name of which may vary with locations, and the picture of the various of rail fastening system are shown as follow:

|  |  | 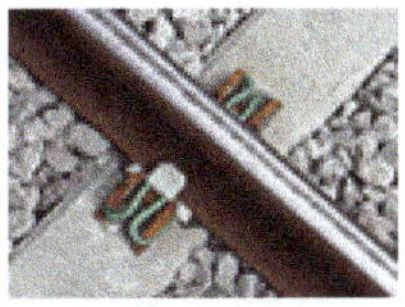 |  |  |  |  |
|---|---|---|---|---|---|---|
| Rail spike with base plate above the tie | E-Clip' fastening | 'fastclip' fastening | Tension clamp fastening | Bolt clamped fastening | Track joint and chairs | Steel spring keyed rail in chair |

## Fastening Design Criteria

A resilient fastening has its own features, such as, the spring/clip holding the rail to the railroad sleeper, the shoulder/anchor holding the clip, an insulator and a rail pad separating the rail from the railroad sleeper. As for the four features, there are strict criteria stipulated to ensure the fastening effect.

## Rail Clamping Force

Clamping forces may vary with the fastening system and customer requirements. Generally, most fastening systems offer a clamping force per clip between 7.5 to 12.5 kN with deflections of the clips toe between 10 and 15mm. According to European standards, the minimum force to push the rail through the fastening system of 7kN is for most main line tracks and 9 kN for high speed rail and heavy freight lines, which suggests that a nominal clamping force per clip of 8.5kN at minimal for mixed traffic line and 10 kN for the more severe applications.

## Shoulder/Anchor

The shoulder/anchor has a very important role in fastening system. The shoulder/anchor has to be able to withstand the impact of loads and vibration transmitted to the sleeper without breaking the sleeper or getting loose. The European standard requires that all the elements of fastening should be capable to withstand a 60kN pullout force with doing no harm to the sleeper.

## Insulator

Insulator in rail fastening system works as a cushion to stop the wearing of other fastening components. And then the insulator must be made of materials resistant to wear, to degrade from ultra violet light and to attack from chemicals. The requirements on insulators vary with the nature of the signaling and electrical systems.

## Rail Pads

The function of rail pads is to spread the load from the rail to the railroad sleeper and to attenuate the loads impact from bad joints, track irregularities and faults in rolling stock. The railroad pad is usually made of rubber, of 10 mm thick, or is a plastic such as EVA between 5 mm and 10 mm thick. The stiffness of the pad is typically in the rang 40-450 kN/mm, and the 5 mm plastic pad can offer stiffness as high up to 6000 kN/mm. Low stiffness railroad pads are not recommended for use in curves sharper than about 400m radius.

# Railroad Switch

A railroad switch (AE), turnout, or [set of] points (BE) is a mechanical installation enabling railway trains to be guided from one track to another, such as at a railway junction or where a spur or siding branches off.

The switch consists of the pair of linked tapering rails, known as *points* (*switch rails* or *point blades*), lying between the diverging outer rails (the *stock rails*). These points can be moved laterally into one of two positions to direct a train coming from the point blades toward the straight path or the diverging path. A train moving from the narrow end toward the point blades (i.e. it will be directed to one of the two paths depending on the position of the points) is said to be executing a *facing-point movement.*

Unless the switch is locked, a train coming from either of the converging directs will pass through the points onto the narrow end, regardless of the position of the points, as the vehicle's wheels will force the points to move. Passage through a switch in this direction is known as a *trailing-point movement.*

A switch generally has a straight "through" track (such as the main-line) and a diverging route. The handedness of the installation is described by the side that the diverging track leaves. *Right-hand switches* have a diverging path to the right of the straight track, when coming from the point blades, and a *left-handed switch* has the diverging track leaving to the opposite side. In many cases, such as rail yards, many switches can be found in a short section of track, sometimes with switches going both to the right and left (although it is better to keep these separated as much as feasible). Sometimes a switch merely divides one track into two; at others, it serves as a connection between two or more parallel tracks, allowing a train to *switch* between them. In many cases, where a switch is supplied to leave a track, a second is supplied to allow the train to reenter the track some distance down the line; this allows the track to serve as a siding, allowing a train to get off the track to allow traffic to pass (this siding can either be a dedicated short length of track, or formed from a section of a second, continuous, parallel line), and also allows trains coming from either direction to switch between lines; otherwise, the only way for a train coming from the opposite direction to use a switch would be to stop, and reverse through the switch onto the other line, and then continue forwards (or stop, if it is being used as a siding).

A *straight track* is not always present; for example, both tracks may curve, one to the left and one to the right (such as for a wye switch), or both tracks may curve, with differing radii, while still in the same direction.

## Operation

A railroad car's wheels are guided along the tracks by coning of the wheels. Only in extreme cases does it rely on the flanges located on the insides of the wheels. When the wheels reach the switch, the wheels are guided along the route determined by which of the two points is connected to the track facing the switch. In the illustration, if the left point is connected, the left wheel will be guided along the rail of that point, and the train will diverge to the right. If the right point is connected, the right wheel's flange will be guided along the rail of that point, and the train will continue along the straight track. Only one of the points may be connected to the facing track at any time; the two points are mechanically locked together to ensure that this is always the case.

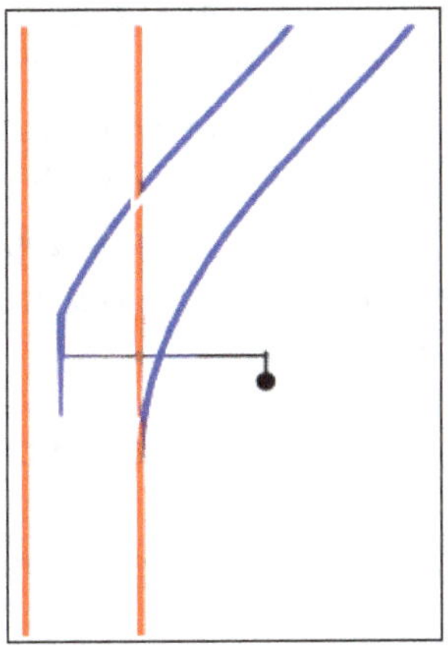

The operation of a railroad switch. The red track is the one travelled during a facing-point movement. The switch mechanism, shown in black, may be operated remotely using an electric motor or hand-operated lever or from a nearby ground frame.

A mechanism is provided to move the points from one position to the other. Historically, this would require a lever to be moved by a human operator, and some switches are still controlled this way. However, most are now operated by a remotely controlled electric motor or by pneumatic or hydraulic actuation, called a point machine. This both allows for remote control and for stiffer, strong switches that would be too difficult to move by hand, yet allow for higher speeds.

In a trailing-point movement (running through the switch in the wrong direction while they are set to turn off the track), the flanges on the wheels will force the points to the proper position. This is sometimes known as *running through the switch*. Some switches are designed to be forced to the proper position without damage. Examples include variable switches, spring switches, and weighted switches.

If a switch becomes worn or the operating rods become damaged, it is possible for the flange to *split the switch*, and go through the switch in the direction other than what was expected. This happens when the flange strikes a small gap between the fixed rail and the set switch point (whichever is touching the main line); this forces the switch open, and the train is diverted down the incorrect track. This can either happen to the locomotive, in which case the whole train can be directed onto the wrong track, with potentially dangerous results, or it can occur at any point through the train, when a random truck is directed down a different track from the rest of the train; if this happens on the front truck of a car, the usual result is derailment, as the trailing truck of the preceding car attempts to go one way, while the leading truck of the following car tries to go another. If it happens to the trailing truck of a car, the front truck will follow one track, while the trailing truck follows a parallel line; this causes the whole car to "crab", or move sideways down the track (derailment often results eventually, due to the lateral forces applied when the train tries to brake or accelerate). This can have disastrous results if there is any obstacle between the lines, as the car will be propelled into it sideways, such as happened in the 1928 Times Square derailment. In some cases, the whole train behind the car will follow the errant car onto the other track; in others, only one or a few trucks are diverted, while the rest follow the correct track. In cases where it is a simple siding, rather than a continuous parallel track, the diverted truck(s) can travel the whole length of the siding until it turns back to the main track, where it performs a *trailing point movement*, forces the switch open, and ends up back on the same track again, with only damage to the switches. This is far less likely in cases of diversion to a parallel track, since switches on both lines will often be interconnected, so to set the switch on the main line to straight-through will set the other switch to straight-through as well (otherwise there is a risk of turning off the track only to find the joining switch is set the wrong way, and running the train through it). Because derailments are expensive and very dangerous to life and limb, maintenance of switch points and other trackwork is essential, especially with faster trains. Another derailment that occurred due to a split switch is the ProRail Hilversum derailment on 15 January 2014.

If the points are rigidly connected to the switch control mechanism, the control mechanism's linkages may be bent, requiring repair before the switch is again usable. For this reason, switches are normally set to the proper position before performing a trailing-point movement.

An example of a mechanism that would require repair after a run-through in the trailing direction is a clamp-lock. This mechanism is popular in the UK, but the damage caused is common to most types of switches.

It would be possible, at least theoretically, to build a rail switch with linkages strong enough that they would not bend under the force of the flanges of train wheels pushing one of the points away from the adjacent fixed rail, so that the points would never move during a trailing-point movement, at least as long as the speed of the train was not excessive. Then, in a trailing-point movement along the route that the points were not set to, the switch would not be damaged, but instead the train would derail. Obviously, it is preferable for the switch to give way and be damaged than for the train to derail, causing damage to it and possible injury or loss of life to people aboard the train or nearby.

## High-speed Operation

Generally, switches are designed to be safely traversed at low speed. However, it is possible to modify the simpler types of switch to allow trains to pass at high speed. More complicated switch systems, such as double slips, are restricted to low-speed operation. On European high speed lines, it is not uncommon to find switches where a speed of 200 km/h (124.3 mph) or more is allowed on the diverging branch. Switches were passed over at a speed of 560 km/h (348.0 mph) (straight) during the French world speed run of April 2007.

The conventional way to increase turnout speeds is to lengthen the turnout and use a shallower frog angle. If the frog angle is so shallow that a fixed frog cannot support a train's wheels, a swing-nose crossing (US: *moveable point frog*) will be used. Higher speeds are possible without lengthening the turnout by using uniformly curved rail and a very low entry angle; however, wider track centers may be needed.

The US Federal Railroad Administration has published the speed limits for higher-speed turnouts with No. 26.5 turnout that has speed limit of 60 miles per hour (96.6 km/h) and No. 32.7 with speed limit of 80 miles per hour (128.7 km/h).

## Operation in Cold Conditions

Gas heating keeps a switch free from snow and ice.

In cold conditions, snow and ice can prevent the correct operation of switches. In the past, people were employed by railway companies to keep the switches clear by sweeping the snow away, and this is still used in some countries, especially on minor lines. Some were provided with gas torches for melting ice. More recently, switches have had heaters installed in the vicinity of the points so that the temperature of the rails in these areas can be kept above freezing. The heaters may be powered by gas or electricity. In cases where gas or electric heaters cannot be used due to logistical or economical constraints, anti-icing chemicals can be applied to create a barrier between the metal surfaces of the switch and ice.

Heating alone may not be enough to keep switches functioning in snowy conditions. If the snow is particularly sticky, as it may be in temperatures just below freezing, chunks of ice may accumulate on trains. When the train passes over some switches, the shocks, possibly in combination with slight heating caused by braking or a city microclimate, may cause the chunks of ice to fall off, jamming the switches. The heaters need time to melt the ice, so if service frequency is high, there may not be enough time to melt the ice before the next train arrives, disrupting services. Possible solutions are installing higher capacity heaters, reducing the frequency of the timetable or applying anti-icing chemicals like ethylene glycol to the trains.

## Tram and Monorail Systems

Monorail.

The traditional solution for control is whether the car draws power or not when passing under a special short segment of the overhead wire. The presence or absence of power draw is detected by special circuitry, which activates or deactivates the switch points. This arrangement requires the tram car to coast unpowered through the switch (running on momentum) when making certain moves.

The next system uses a powerful electromagnet in the tram and a reed relay inlaid between the tracks to initiate the blade turning mechanism. The driver has a separate switch to control the magnet, thus switching is no longer dependent on power draw of the tram simplifying the procedure somewhat. Turning the switchblades always requires a magnetic field on the relay and its polarity dictates the direction. No magnetic field when passing the relay means retain the blades in whatever the position they were. Alternatively, more recently, radio telemetry or some other form of control signaling is used. Monorail systems have special switches.

## Roller Coaster Switches

Substitution track switch for rail at Chester Zoo.

Many roller coasters have switches for the siding, or even for a double station system, for example in Disneyland Paris' Space Mountain and Galactica at Alton Towers.

Regular rail can cross its own track because the gaps in the rails for wheel flanges are narrow, permitting the bladed design. Round pipe roller coaster rails and box beam monorail rails usually have wheels riding at angles other than on top. These additional other angle wheels are a larger loading gauge, requiring big gaps in the rail (structure gauge) where rails cross or meet.

There are three basic switch designs for roller coasters. Flexing, substituting and table rotating rails have all been used. Flexing the entire rail truss, fixed at one end, to point towards an alternate destination requires manipulating a long segment of rail. Substituting a segment requires placing two or more segments of rail on flat plate that is moved in its entirety to provide straight or curved track. Alternatively these substitution track segments can be wrapped around a rotating cylinder, creating a triangular truss or a two sided plate. Rotating a table with a curved track segment in a Y junction is the less used third option. If the curved track turns the cars 60 degrees, and three rail lines meet as three equally spaced spokes, 120 degrees apart, then the curved track sitting on a turn table can be rotated to connect any two of the three rail lines at this junction, creating a triangle junction.

## Classification

The divergence and length of a switch is determined by the angle of the frog and the angle or curvature of the switch blades. The length and placement of the other components are determined from this using established formulas and standards. This divergence is measured as the number of units of length for a single unit of separation.

A low-speed right-hand main line to yard switch.

In North America this is generally referred to as a switch's "number". For example, on a "number 12" switch, the rails are one unit apart at a distance of twelve units from the center of the frog.

In the United Kingdom points and crossings using chaired bullhead rail would be referred to using a letter and number combination. The letter would define the length (and hence the radius) of the switch blades and the number would define the angle of the crossing (frog). Thus an A7 turnout would be very short and likely only to be found in tight places like dockyards whereas an E12 would be found as a fairly high speed turnout on a mainline.

## Safety

Switches are crucial to the safe running of a railway because they pose a number of risks:

- Incorrectly set points may result in two trains being on the same track, potentially causing a collision. The wrong setting might be caused by tampering with a manually operable switch or by operation errors in an interlocking.
- Reversing the points under a moving train will almost always derail the train.
- Points might move due to the extreme forces exerted by a passing train.
- A train might stand so close to the frog of a switch that a passing train might collide with its side (the first train is then said to *foul the switch*).
- The necessary maintenance of the complex mechanical device might be neglected.

To avoid accidents caused by these risks, suitable technical remedies as well as certain practices are applied. The most important are:

- Locks to prevent reversing a switch without a proper key.
- Interlockings that allow signals only to be cleared when switches are set correctly.
- Track circuits to prevent reversing when a passing train is detected.
- Point locks or clamps that prevent movement of blades by locking them to the stock rails in a secure manner.
- Track circuits and fouling markers to point out fouling vehicles.
- Maintenance schedules, especially for measuring deviations of critical distances.

Switch-related accidents caused by one or more of these risks have occurred, including:

- The 1980 Buttevant Rail Disaster at Buttevant, County Cork, in Ireland, when the Dublin–Cork express was derailed at high speed after being inadvertently switched into a siding via ground frame operated points, resulting in 18 deaths.
- Wrecks caused by switches being thrown open in front of the trains by saboteurs, as in the non-fatal derailments near Newport News, Virginia, on August 12, 1992, and in Stewiacke, Nova Scotia, on April 12, 2001. To prevent these incidents, most unused switches are locked up.
- The 1998 Eschede train disaster in Germany was one of the world's deadliest high-speed train accidents, resulting in over 100 deaths. It occurred when a wheel rim failed at 200 km/h (125 mph), partially derailing the car. The wheel rim went through the floor of the carriage and was dragging on the ground. On arrival at the junction it threw the switch, causing the rear wheels of the car to switch onto a track parallel to the track taken by the front wheels. The car was thereby thrown into and destroyed the piers supporting a 300-tonne roadway overpass.
- The May 2002 Potters Bar rail crash at Potters Bar, Hertfordshire, in the United Kingdom, occurred when a switch sprang to a different position as a coach crossed it, a type of mishap called *splitting the switch*. The front wheels of a coach progressed along the straight track as intended, but the rear wheels slewed along the diverging track. This caused the whole coach to detach from the train and slew sideways across the platform ahead. The movement of the switch occurred beneath the final coach, so that although seven people were killed, the front coaches remained on the tracks. Poor maintenance of the points was found to be the primary cause of the crash.
- The initial conclusion of the inquiry into the Grayrigg derailment of February 23, 2007, blames an incorrectly maintained set of points.
- Two fatal accidents involving the Amtrak *Silver Star* passenger train in South Carolina caused by switches either malfunctioning or misaligned. On July 31, 1991, several cars derailed killing 7 passengers, due to a missing locking pin on the switch mechanism. Seventeen years later, on August 4, 2018, the Silver Star crashed into a parked freight train on a siding due to a misaligned switch, killing two crewmen.

## Components

This detail of a switch shows the pair of tapered moveable rails known as the *switch points* (*switch rails* or *point blades*).

## Points (Point Blades)

The *points* (*switch rails* or *point blades*) are the movable rails which guide the wheels towards either the straight or the diverging track. They are tapered on most switches, but on *stub switches* they have square ends.

In the UK and Commonwealth countries, the term *points* refers to the entire mechanism, whereas in North America the term refers only to the movable rails.

In some cases, the switch blades can be heat treated for improvement of their service life. There are different kinds of heat treatment processes such as edge hardening or complete hardening.

## Frog (Common Crossing)

A one-piece cast *frog*. The shiny line crosses the rusty line. This North American "self-guarding cast manganese" frog without guard rails has raised flanges on the frog, bearing on the face of the wheel as it passes through the frog.

The *frog* and *guard rail* of a switch.

The frog, also known as the common crossing (or V-rail in Australian terminology), is the crossing point of two rails. This can be assembled out of several appropriately cut and bent pieces of rail or can be a single casting of manganese steel. On lines with heavy use the casting may be treated with explosive shock hardening to increase service life.

On lines with heavy or high-speed traffic, a swingnose crossing (movable-point frog) may be used. As the name implies, there is a second mechanism located at the frog. This moves a small portion of rail, to eliminate the gap in the rail that normally occurs at the frog. A separate switch machine is required to operate the movable-point frog switch.

This term *frog* is taken from the part of a horse's hoof it most closely resembles. Certain types of overhead electrification systems that make use of trolley poles have similar devices referred to as wire frogs.

On dual-gauge switches, a special frog is used where the third rail crosses the common rail. Denver and Rio Grande crews called this a "toad".

A recent development on North American freight railroads is the flange-bearing frog, in which the wheel flange supports the weight of the vehicle as opposed to the tread. This design reduces impact loading and extends the life of the frog.

## Guard Rail (Check Rail)

A *guard rail* (*check rail*) is a short piece of rail placed alongside the main (stock) rail opposite the frog. These ensure that the wheels follow the appropriate flangeway through the frog and that the train does not derail. Generally, there are two of these for each frog, one by each outer rail. Guard rails are not required with a "self-guarding cast manganese" frog, as the raised parts of the casting serve the same purpose.

*Check rails* are often used on very sharp curves, even where there are no switches.

The *switch motor* (in this case an electric motor) and associated mechanism used to operate this switch can be seen to the in the picture.

## Switch Motor

A *switch motor* (also known as a switch machine, point motor or point machine) is an electric, hydraulic or pneumatic mechanism that aligns the points with one of the possible routes. The motor is usually controlled remotely by the dispatcher (signaller in the UK). The switch motor also includes electrical contacts to detect that the switch has completely set and locked. If the switch fails to do this, the governing signal is kept at red (stop). There is also usually some kind of manual handle for operating the switch in emergencies, such as power failures.

A patent by W. B. Purvis dates from 1897.

An example of a mechanism used at a switch. The two points are linked together with a throw bar. The throw bar extends to the lever on the near side of the track, which is used to *throw the switch*.

This is an example of a low switch stand, used at locations where there is not sufficient clearance for a tall switch stand. This particular stand is designed to be trailed through by rolling stock, which will cause the points to become lined for the route that the wheels have passed through. It has a reflectorised target.

## Points Lever

A *points lever*, *ground throw*, or *switchstand* is a lever and accompanying linkages that are used to align the points of a switch manually. This lever and its accompanying hardware is usually mounted to a pair of long sleepers that extend from the switch at the points. They are often used in a place of a switch motor on infrequently used switches. In some places, the lever may be some distance from the points, as part of a lever frame or ground frame. To prevent the tampering of switches by outside means, these switches are locked up when not in use.

## Point Machine Conversion

A point machine conversion system consist in a remotely controlled device attached to an existing manually operated point that allows the shunter or driver to remotely operate hand points with a radio handset. Each converter can be used as a stand-alone or multiple units can be installed operating together with routing.

## Facing Point Lock

A set of points on the Strathspey Railway in Scotland. The facing point lock in the middle will need to be withdrawn using one of the two levers on the left before the points themselves can be moved using the other lever. Once the points have been moved the lock will be pushed in again to lock the points in position.

A *facing point lock* (*FPL*), or *point lock*, is a device which, as the name implies, locks a set of points in position, as well as proving that they are in the correct position. The *facing point* part of the name refers to the fact that they are to prevent movement of the points during facing moves, where a train could potentially *split* the points (end up going down both tracks) if the points were to move underneath the train – during trailing moves, the wheels of a train will force the points into the correct position if they attempt to move.

In the United Kingdom, FPLs were common from an early date, due to laws being passed which forced the provision of FPLs for any routes travelled by passenger trains – it was, and still is, illegal for a passenger train to make a facing move over points without them being locked, either by a point lock, or temporarily clamped in one position or another.

## Joints

*Joints* are used where the moving points meet the fixed rails of the switch. They allow the points to hinge easily between their positions. Originally the movable switch blades were connected to the fixed closure rails with loose joints, but since steel rails are somewhat flexible it is possible to make this join by thinning a short section of the rail itself. This can be called a heelless switch.

## Straight and Curved Switches

Turnouts were originally built with straight switch blades, which ended at the pointed end with a sharp angle. These switches cause a bump when the train traverses in the turnout direction. The switch blades could be made with a curved point which meets the stockrail at a tangent, causing less of a bump, but the disadvantage is that the metal at the point is thin and necessarily weak. A solution to these conflicting requirements was found in the 1920s on the German Reichsbahn. The first step was to have different rail profile for the stock rails and switch rails, with the switch rails being about 25 mm (0.98 in) less high, and stockier in the middle.

## Point Indicators

As it is difficult to see the lie of a switch from a distance, especially at night, European railways and their subsidiaries provide point indicators which are often illuminated.

A swingnose crossing: The point of the V-shaped rail is moved to align the rail in the appropriate direction where the two rails cross.

Several different styles of switch stand on display at the Mid Continent Railway Museum.

A ground frame contains a few levers for manually operating nearby points: Blue lever: Release, Black lever: Points, Red lever: Signal.

A light industrial or yard track switch joint, where the points are joined to the closure rails by bolts through a "joint bar" or "fish plates".

Railroad switch of a narrow gauge railway track.

## Types

Apart from the standard right-hand and left-hand switches, switches commonly come in various combinations of configurations.

### Slip Switches

### Double Slip

A double switch, or double slip. The points are set to connect the upper left and lower right tracks.

A *double slip switch* (*double slip*) is a narrow-angled diagonal flat crossing of two lines combined with four pairs of points in such a way as to allow vehicles to change from one straight track to the other, as well as going straight across. A train approaching the arrangement may leave by either of the two tracks on the opposite side of the crossing. To reach the third possible exit, the train must change tracks on the slip and then reverse.

The arrangement gives the possibility of setting four routes, but because only one route can be traversed at a time, the four blades at each end of the crossing are often connected to move in unison, so the crossing can be worked by just two levers or point motors. This gives the same functionality of two points placed end to end. These compact (albeit complex) switches usually are found only in locations where space is limited, such as station throats (i.e. approaches) where a few main lines spread out to reach any of numerous platform tracks.

In North American English, the arrangement may also be called a *double switch*, or more colloquially, a *puzzle switch*. The Great Western Railway in the United Kingdom used the term *double compound points*, and the switch is also known as a *double compound* in Victoria (Australia). In Italian, the term for a double switch is *deviatoio inglese*, which means *English switch*. Likewise, it is called *Engels(e) Wissel* in Dutch, and was called *Engländer* in German in former times.

## Single Slip

A *single slip switch* works on the same principle as a double slip, but provides for only one switching possibility. Trains approaching on one of the two crossing tracks can either continue over the crossing, or switch tracks to the other line. However, trains from the other track can only continue over the crossing, and cannot switch tracks. This is normally used to allow access to sidings and improve safety by avoiding having switch blades facing the usual direction of traffic. To reach the sidings from what would be a facing direction, trains must continue over the crossing, then reverse along the curved route (usually onto the other line of a double track) and can then move forward over the crossing into the siding.

## Outside Slip

A double, outside slip in Heidelberg main station.

An *outside slip switch* is similar to the double or single slip switches described above, except that the switch blades are outside of the diamond instead of inside. An advantage over an inside slip switch is that trains can pass the slips with higher speeds. A disadvantage over an inside slip switch is that they are longer and need more space.

An outside slip switch can be so long that its slips do not overlap at all, as in the example pictured. In such a case a single, outside slip switch is the same as two regular switches and a regular crossing. An outside, double slip switch is about the same as a *scissors crossover* but with the disadvantages:

- The two parallel tracks cannot be used at the same time.
- That the slips are not straight and thus have a limited speed.

Advantage:

- The crossing can be passed at full speed.

Due to the disadvantages over both the *double inside slip switch* and the *scissors crossover*, double outside slip switches are only used in rare, specific cases.

## Crossover

A scissors crossover: Two pairs of switches linking two tracks to each other in both directions.

A crossover is a pair of switches that connects two parallel rail tracks, allowing a train on one track to cross over to the other. Like the switches themselves, crossovers can be described as either facing or trailing.

When two crossovers are present in opposite directions, one after the other, the four-switch configuration is called a double crossover. If the crossovers in different directions overlap to form an ×, it is dubbed a scissors crossover, scissors crossing, or just scissors; or, due to the diamond in the center, a diamond crossover. This makes for a very compact track layout at the expense of using a level junction.

In a setup where each of the two tracks normally carries trains of only one direction, a crossover can be used either to detour "wrong-rail" around an obstruction or to reverse direction. A crossover can also join two tracks of the same direction, possibly a pair of local and express tracks, and allow trains to switch from one to the other.

On a crowded system, routine use of crossovers (or switches in general) will reduce throughput, as the switches must be changed for each train. For this reason, on some high-capacity rapid transit systems, crossovers between local and express tracks are not used during normal rush hour service, and service patterns are planned around use of the usually flying junctions at each end of the local-express line.

Überleitstelle (crossover) at Richthof between Kirchheim and Langenschwarz stations on the Hanover–Würzburg high-speed railway.

In German a crossover is known as an Überleitstelle (abbreviated to Üst) and is defined as an operating control point on the open line. It is also a block section. At an Überleitstelle trains can transfer from one track of a single or double track section of route to another track on a double track section on the same route. Depending on the safety equipment provided, trains may run this other track either by exception or routinely against the normal direction of traffic.

An Überleitstelle must have at least one turnout. On double tracked routes, single and double crossovers are common, each one consisting of two turnouts and an intermediate section. Very often – but not mandatory – the turnouts and block signals at an Überleitstelle are remotely controlled or set from a central signal box.

The official categorisation of an Überleitstelle as a type of junction first arose in Germany with the construction of high-speed railways. Previous to that there were already operating control points at which trains could just transfer from one track to another on the same route, but they were considered as junctions (Abzweigstelle). The latter are still used to refer to those places in stations which enable trains to cross from one route to another.

## Stub Switch

Closeup of a stub switch.

A narrow gauge *stub switch*. Note that this switch has an additional piece of movable rail instead of a frog.

A *stub switch* lacks the tapered points (point blades) of a typical switch. Instead, both the movable rails and the ends of the rails of the diverging routes have their ends cut off square. The switch mechanism aligns the movable rails with the rails of one of the diverging routes. In 19th century US railroad use, the stub switch was typically used in conjunction with a harp switch stand.

The rails leading up to a stub switch are not secured to the sleepers for several feet, and rail alignment across the gap is not positively enforced. Stub switches also require some flexibility in the rails (meaning lighter rails), or an extra joint at which they hinge. Therefore, these switches cannot be traversed at high speed or by heavy traffic and so are not suitable for main line use. A further disadvantage is that a stub switch being approached from the diverging route that is not connected by the points would result in a derailment. Yet another disadvantage is that in very hot weather, expansion of the steel in the rails can cause the movable rails to stick to the stock rails, making switching impossible until the rails have cooled and contracted.

One advantage to stub switches is that they work better in the snow. The sideways action of the point rails pushes snow to the side, instead of packing the snow between the points and the rail in a more modern design.

Stub switches were more common in the very early days of railways and their tramway predecessors. Now, because of their disadvantages, stub switches are used primarily on narrow gauge lines and branch lines. Some modern monorail switches use the same principle.

## Plate Switch

A narrow gauge *plate switch*.

A *plate switch* incorporates the tapered points of a typical switch into a self-contained plate. Each point blade is moved separately by hand. Plate switches are only used for double-flanged wheels, with wheels running through the plates on their flanges, guided by the edges of the plate and the moveable blade.

Because plate switches can only be used by double-flanged wheels and at extremely low speeds, they are typically only found on hand-worked narrow gauge lines.

## Three-way Switch

A three-way stub switch at Sheepscot station on the Wiscasset, Waterville and Farmington Railway.

A *three-way switch* is used to split a railroad track into three divergent paths rather than the more usual two. There are two types of three-way switches. In a *symmetrical three-way switch*, the left and right branches diverge at the same place. In an *asymmetrical three-way switch*, the branches diverge in a staggered way. Both types of three-way switches require three frogs.

The complexity of symmetrical switches usually results in speed restrictions, therefore three-way switches are most often used in stations or depots where space is restricted and low speeds are normal. Symmetrical switches were used quite often on Swiss narrow-gauge railways. Asymmetrical three-way switches are more common, because they do not have speed restrictions compared to standard switches. However, because of their higher maintenance cost due to special parts as well as asymmetric wear, both types of three-way switches are replaced with two standard switches wherever possible.

In areas with very low speeds, like depots, and on railroads that had to be built very cheaply, like logging railroads, three-way switches were sometimes built as stub switches.

## Off-railer

The off-railer is a system of installing a turnout over and above some plain track, without having to cut or replace that track. It is useful for installing temporary branches on agricultural railways, and sidings for track machines on mainline rails. Special ramps lift the wheels off the normal track, and then the off-railer curves away as required. Decauville has such a system. It is a bit like a drawbridge crossing.

## Interlaced Turnout

Chicago Transit Authority switch tower 18 interlaced turnout.

Interlaced turnouts on the elevated Chicago "L" north and southbound Purple and Brown lines intersecting with east and westbound Pink and Green lines and the looping Orange line above the Wells and Lake street intersection in The Loop.

An *interlaced turnout* is a different method of splitting a track into three divergent paths. It is an arrangement of two standard turnouts, one left- and one right-handed, in an "interlaced" fashion.

The points of the second turnout are positioned between the points and the frog of the first turnout. In common with other forms of three way turnouts an additional common-crossing is required. Due to the inherent complexity of the arrangement, interlaced turnouts are normally only used in locations where space is exceptionally tight, such as station throats or industrial areas within large cities. Interlaced turnouts can also be found in some yards, where a series of switches branching off to the same side are placed so close together that the points of one switch are placed before the frog of the preceding switch.

## Wye Switch

A *wye switch.*

A *wye switch* (*Y points*) has trailing ends which diverge symmetrically and in opposite directions. The name originates from the similarity of their shape to that of the letter Y. Wye switches are usually used where space is at a premium. In North America this is also called an "equilateral switch" or "equilateral turnout". Common switches are more often associated with mainline speeds, whereas wye switches are generally low-speed yard switches.

One advantage of wye switches is that they can have a coarser frog angle using the same radius of curvature than a common switch. This means that they give rise to a less severe speed restriction than the diverging branch of a common switch, without having to resort to more expensive switches with a moving frog. For this reason they are sometimes used on a main line where it splits into two equally important branches or at the ends of a single track section in an otherwise double track line.

## Run-off Points

Trap points at the exit from a yard.

*Run-off points* are used to protect main lines from stray or runaway cars, or from trains passing signals set at danger. In these cases, vehicles would otherwise roll onto and foul (obstruct) the main line and cause a collision. Depending on the situation in which they are used, run-off points are referred to either as trap points or catch points. Derailers are another device used for the same purpose.

*Catch points* are installed on the running line itself, where the railway climbs at a steep gradient. They are used to prevent runaway vehicles colliding with another train further down the slope. In some cases, catch points lead into a sand drag to safely stop the runaway vehicle, which may be travelling at speed. Catch points are usually held in the 'derail' position by a spring. They can be set to allow a train to pass safely in the downhill direction using a lever or other mechanism to override the spring for a short time.

Catch points originate from the days of the 'unfitted' goods (freight) train. As these trains tended to consist of either completely unbraked wagons (relying entirely on the locomotive's own brakes), or ones with unlinked, manually applied brakes (necessitating a stop at the top of steep downgrades for the guard to walk along the train and set the brakes on each wagon in turn), they also lacked any mechanism to automatically brake runaway cars. Catch points were therefore required to stop the rear portion of a poorly coupled train that might break away whilst *climbing* a steep grade – although they would also stop vehicles that ran away for any other reason. Now that trains are all 'fitted' (and broken couplings are far less common), catch points are mostly obsolete.

Similar to catch points, *trap points* are provided at the exit from a siding or where a goods line joins a line that may be used by passenger trains. Unless they have been specifically set to allow traffic to pass onto the main line, the trap points will direct any approaching vehicle away from the main line. This may simply result in the vehicle being derailed, but in some cases a sand drag is used, especially where the vehicle is likely to be a runaway travelling at speed due to a slope.

## Derailers

A *derailer* works by derailing any vehicle passing over it. There are different types of derailers, but in some cases they consist of a single switch point installed in a track. The point can be pulled into a position to derail any equipment that is not supposed to pass.

## Dual Gauge Switches

A dual gauge switch 2005.

*Dual gauge switches* are used in dual gauge systems. There are various possible scenarios involving the routes that trains on each gauge may take, including the two gauges separating or one gauge being able to choose between diverging paths and the other not. Because of the extra track involved, dual gauge switches have more points and frogs than their single gauge counterparts. This limits speeds even more than usual.

A related formation is the 'swish' or rail exchange, where (usually) the common rail changes sides. These have no moving parts, the narrower gauge wheels being guided by guard rails as they transition from one rail to another. The wider gauge only encounters continuous rail so is unaffected by the exchange. At dual gauge turntables, a similar arrangement is used to move the narrow gauge track from one side to a central position.

## Rack Railway Switches

Railroad switch on a rack railway.

Rack railway switches are as varied as rack railway technologies. Where use of the rack is optional, as on the Zentralbahn in Switzerland or the West Coast Wilderness Railway in Tasmania, it is common to place turnouts only in relatively flat areas where the rack is not needed. On systems where only the pinion is driven and the conventional rail wheels are idlers, such as the Dolderbahn in Zürich, Štrbské Pleso in Slovakia and the Schynige Platte rack railway, the rack must be continuous through the switch. The Dolderbahn switch works by bending all three rails, an operation that is performed every trip as the two trains pass in the middle. The Štrbské Pleso and Schynige Platte Strub rack system instead relies on a complex set of moving points which assemble the rack in the traversed direction and simultaneously clear the crossed direction conventional rails. In some rack systems, such as the Morgan system, where locomotives always have multiple driving pinions, it is possible to simplify turnouts by interrupting the rack rail, so long as the interruption is shorter than the spacing between the drive pinions on the locomotives.

Some systems use transfer tables instead to provide continuous rack. The Pilatus Railway has unusual switches that rotate on an axis parallel to the track.

## Switch Diamond

A switch diamond at a junction.

Although not strictly speaking a turnout, a *switch diamond* is an active trackwork assembly used where the crossing angle between two tracks is too shallow for totally passive trackwork: the unguided sections of each rail would overlap. These vaguely resemble two standard points assembled very closely toe-to-toe. These would also often utilise swingnose crossings at the outer ends to ensure complete wheel support in the same way as provided on shallow angle turnouts. In North America these are known as *Movable-Point Diamonds*. In the UK, where the angle of divergence is shallower than 1 in 8 (center-line measure) a switched diamond will be found rather than a passive or fixed diamond.

Such switches are usually implemented on the basis of increasing the safe crossing speed. Open blades impose a speed restriction, due to the potential of the crossing impact fracturing the rail as both wheels on each axle hit the crossing gaps almost simultaneously. Switched blades, as shown in the photograph, allow a much higher speed across the gap by providing an essentially continuous piece of rail across the gap on both sides.

The frog end of the switched crossing, despite still having a gap in one rail, is less problematic in this regard. The outer rail is still continuous, the wing rail (the part that turns out, after the frog gap) provides a gradual transition, and the check rail avoids the possibility of points splitting. This can be seen in how, under examination, the wing rail has a wider polished section, showing how the wheel load is transferred across the gap.

## Single-point Switch

A single-point switch on the Toronto streetcar system.

Single point switches, known as Tongue and Plain Mate switches, are sometimes used on freight railways in slow speed operation in paved areas such as in ports. In the United States, they are regulated by provision 213.135(i) of the Federal Railroad Administration Track Safety Standards.

On streetcar (tram) systems using grooved rails, if the wheels on both sides of the car are connected by a rigid solid axle, only one switchpoint is needed to steer it onto one or the other track. The switchpoint will be on inside rail of the switch's curve route. When a streetcar enters the curve route of the switch, the wheel on the inside of the curve (the right side of the car on a right turn) is pulled into the turn, and through the axle, directs the wheel on the outside to also follow the curve. The outside wheel is supported for a short distance by its flange running in the groove.

Some low floor streetcar designs use split axles (a separate half-axle for the wheel on each side of the car). Such streetcars are unsuitable for use with single-point switches as there would be no mechanism to transfer the force from the inner to outer wheels at switches.

A single-point switch is cheaper to build, especially in street trackage, as there is no need to link to a second switchpoint.

## Expansion Joint

Expansion joints look like a part of a railroad switch, but have a completely different purpose, namely to compensate for the shrinkage or expansion of the road bed - e.g. typically, a larger steel bridge - due to changes in temperature.

## Turnout Speeds

A Railroad Switch in Wazir Mansion Station.

Turnout speeds are governed by a number of factors.

As a general rule, the smaller the crossing angle of a turnout, the higher the turnout speed. In North America, turnouts are rated numerically, which represents the ratio of divergence per length as measured at the frog. A rule of thumb is that the rated speed of a switch (in miles per hour) is twice the numerical rating:

- No. 15: 30 mph (48 km/h).
- No. 20: 40 mph (64 km/h).

Higher speed turnouts have also been used in the United States:

- No. 26.5: 60 mph (97 km/h).
- No. 32.7: 80 mph (130 km/h).

In most other countries, switches are marked with tangent of crossing angle. For example, Russia and the rest of the Commonwealth of Independent States (CIS) use the following designations:

- 1/6: sorting yards only, whenever it is impossible to install a better switch.
- 1/9: 40 km/h (25 mph), the most common switch, installed by default.

- 1/11: 50 km/h (31 mph), used where passenger trains follow a diverging path. Swingnose crossing may be installed if required.
- 1/18: 80 km/h (50 mph), used where either non-interruptible movement is required or the mainline diverges from the branch line.
- 1/22: 120 km/h (75 mph), rarely used, high-speed lines only.

In Germany, Austria, Switzerland, Poland and other European countries, switches are described by the radius of the branching track (in meters) and the tangent of the frog angle. The crossing may be straight, as in a crossover, or curved for other uses. The following designations are typical examples:

- 190-1:9, the most common switch, for 40 km/h on the branch track.
- 300-1:9, rarely used, for 50 km/h.
- 300-1:12, for 60 km/h.
- 1200-1:18.5, for 100 km/h.
- 2500-1:26.5, for 130 km/h.

Other considerations include the type of turnout (e.g. normal nose, swing nose, slips), wear and tear issues, and the weight and type of the vehicle passing over. Speeds for a trailing movement may be higher than for a facing movement. In many systems, speed limits vary depending on the type of train; for example, a turnout can have a "normal" speed limit for locomotive hauled trains, and a higher speed for multiple unit or high speed trains.

Turnouts with curved or tangential switch blades have higher speed than old style turnouts with straight switch blades.

Older turnouts use the same rail section, shaved down, for both stock rail and switch blade. Newer tangential turnouts use a stubbier rail section for the switch blade.

## Tangential Turnouts

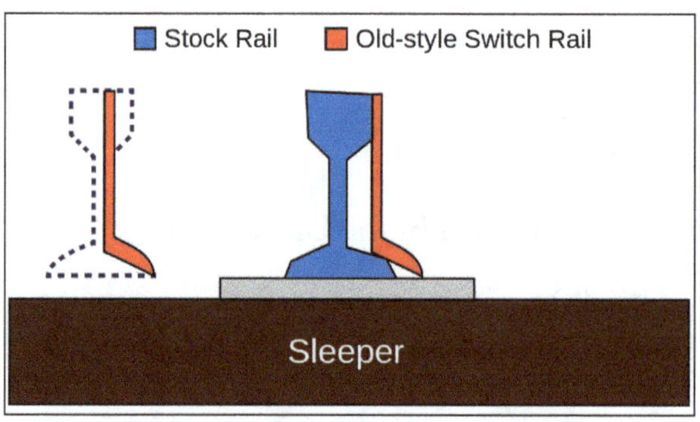


A lot of steel shaved away; tips are weak and straight leading to slow speeds on turnout.

For turnouts, the tips of the switch blades have to be planed down to fit snugly against the fixed or *stock* rail. The diagram shows this: on the left is the dotted outline of the rail, with the part remaining after planing shown by a solid line in red. On the right is shown how the two rails fit together when the turnout is closed. The resulting thin pieces of steel are weak, and so old-style turnouts used to make a relatively sudden and sharp angle against the stock rail. With a sudden change in direction, trains were given a jolt and had to proceed slowly.

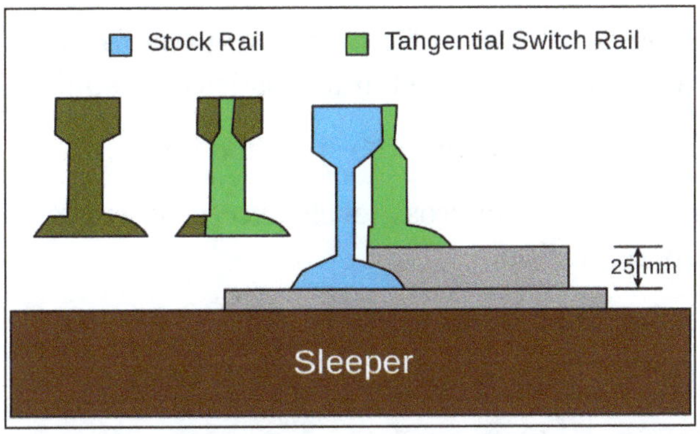


Not so much steel shaved away; tips are strong allowing curved turnouts and higher speeds. Tangential switch (green) rests on chair that is higher.

The diagram shows a how a tangential turnout is made. A lower, more squat profile rail is used to make the switch blades. On the left is the profile of this squat switch rail. The centre diagram shows how the switch rail has to be planed, and on the right it is shown fitting against the stock rail. Note that a thicker base is used to raise the tops level with each other.

The tangential switch rail has less steel planed away, and the mid part of the switch "scallops" into the web of the stock rail for greater strength. The higher baseplate also supports the switch rail better. This makes a stronger switch which can be curved, reducing the jolt to the train and allowing higher speeds. However tangential turnouts still lack a smooth transition where the switch blade contacts the main rail, so there is still some jolt as the train passes over, though it is a much smaller jolt than with the old-style turnouts. The weight of the two types of rail is about the same.

## Assembly and Transport

Transport of switches by rail creates problems as they are so long and wide.

Turnouts are large pieces of rail infrastructure which may be too big, wide, or heavy to transport in one piece. Special wagons can carry the pieces at approximately 45° from vertical, so that they fit within the structure gauge. Once all the pieces have arrived, the turnout is assembled sleeper by sleeper on site. A set of turnouts may be trial assembled beforehand off site, to check that everything fits.

# Essentials of Track Maintenance

A well-maintained track provides the base for a safe and comfortable journey. Therefore, for a track to serve its purpose well, the following characteristics are required of it:

a) The gauge should be correct or within the specified limits.

b) There should be no difference in cross levels except on curves, where cross levels vary in order to provide superelevation.

c) Longitudinal levels should be uniform.

d) The alignment should be straight and kink-free.

e) The ballast should be adequate and the sleepers should be well packed.

f) There should be no excessive wear and tear of the track and all its components and fittings should be complete.

g) Track drainage should be good and the formation should be well maintained. To achieve these standards, the major maintenance operations performed in a calendar year are described.

## Through Packing

Opening of road the ballast is dug out on either side of the rail seat for a depth of 50 mm (2") below the bottom of the sleeper with the help of a shovel with a wire claw. On the outside, the width of the opening should extend up to the end of the sleeper. On the inside, it should extend from the rail seat to a distance of 450 mm (18") in the case of BG, 350 mm (14") in the case of MG, and 250 mm (10") in the case of NG.

Examination of rails, sleepers, and fastenings the rails, sleepers, and fastenings to be used are thoroughly examined. Defective sleepers are removed and loose fastenings are tightened. Any kinks in the rails are removed using a Jim Crow.

Squaring of sleepers the sleepers get out of square quite frequently resulting in gauge variations and kinks. To avoid this, one of the rails is taken as the sighting rail and the correct sleeper spacing is marked on it. The position of the sleeper is checked with reference to the second rail with the help of a T-square. The sleepers are attended to after their defects have been established, which may include their being out of square or at incorrect spacing.

Aligning the track the alignment of the track is normally checked visually, wherein the rail is visually assessed from a distance of about four rail lengths or so. Small errors in the alignment are corrected by slewing the track after loosening the cores at the ends and drawing out sufficient ballast at the end of the sleepers. Slewing is carried out by about six people by planting crowbars deep into the ballast at an angle not exceeding 30 ° from the vertical.

Gauging the gauge should be checked and an attempt should be made to provide a uniform gauge within permissible tolerance limits. Table lists the tolerances prescribed for gauge variation, keeping in mind the side wear that occurs at the time of laying of tracks. This is done to ensure a comfortable ride for the passengers, provided that uniform gauge can be maintained over long lengths.

Table: Gauge tolerance for different tracks.

| Type of track | Gauge tolerance for BG | Gauge tolerance for MG and NG |
|---|---|---|
| Straight track | -6 mm to +6 mm | -3 mm to +6 mm |
| On curves with radius more than 400 m for BG, 290 m for MG, and 175 m for NG | -6 mm to +15 mm | -6 mm to +15 mm |
| On curves with radius less than 400 m for BG, 290 m for MG, and 175 m for NG | Up to + 20 mm | Up to + 20 mm |

The gauge is adjusted in accordance with the type of sleeper under consideration as described in the following:

- Wooden sleepers: In the case of wooden sleepers, gauge adjustment is possible only by removing the dog spikes and refitting them at a new location. Therefore, gauge adjustment should be avoided as far as possible unless the gauge is quiet irregular. When the gauge must be adjusted, all the spikes on the inside and half of those on the outside are removed while the remaining half are loosened. The old spike holes are plugged and new holes are bored in correct places. The gauge on each sleeper is adjusted and the spikes are re-driven.
- Steel trough sleepers: In the case of steel trough sleepers, gauge adjustment is done with the help of keys. When the gauge is slack, the keys on the inside are loosened while those on the outside are driven. The procedure is reversed when the gauge is tight. The maximum possible adjustment of the gauge is 2.5 mm to +4.0 mm.
- CST-9 sleepers: In the case of CST-9 sleepers, gauge adjustment is done with the help of cotters. Normally a gauge is adjusted by 5 mm. The maximum extent to which a gauge can be adjusted -3 mm to +10 mm. It has been noticed that adjusting the gauge may sometimes disturb the alignment, which is taken care of prior to gauging as per the standard practice. In such cases, the track has to be realigned once gauging is completed.

## Packing of Sleepers

The base rail is identified by the mate and the dip or low joints are lifted correctly to ensure that the longitudinal level of the rail is perfect. The sleepers are then packed by applying the scissors packing method. Four men tackle one sleeper simultaneously, two at each rail. The ballast under the sleeper bed is properly packed by the men who stand back to back and work their beaters diagonally by lifting them up to chest level. While the packing is being carried out, the second rail is brought to the correct cross levels thereby ensuring perfect surfacing of the track. In the case of wooden and steel trough sleepers, it should be ensured that the sleepers are not centre-bound and that as such the trough is made in the ballast section at the centre of the sleepers. After packing is completed, the alignment and top should be checked carefully and minor adjustments made as needed.

Repacking of joint sleepers The joint sleepers are then packed once again and the cross levels checked.

Boxing ballast section and dressing Afterwards the ballast section is properly boxed and dressed with the help of a special template. The cess should also be dressed or covered similarly and its level maintained in a way that proper drainage is ensured. Through packing follows a programme which requires that it is undertaken after the monsoon and that it extends from one end of the section to the other. Through packing must be carried out at least once every year.

A gangman normally accomplishes about 11 m to 12 m of through packing on BG, 16 m to 17 m on MG, and 23 m to 24 m on NG tracks.

## Systematic Overhauling

The track should be overhauled periodically with the object of ensuring that the best possible standards of track conditions are met and maintained. The systematic overhauling of the track should normally commence after the completion of one cycle of through packing. It involves the following operations in sequence:

a) Shallow screening and making up of ballast section.

b) Replacement of damaged or broken fittings.

c) All items included in through packing.

d) Making up the cess.

The frequency of overhauling depends upon a number of factors such as the type and age of track structure, the maximum permissible speed and volume of traffic, the mode of traffic, the mode of traction, the rate of track deterioration, and the amount of rainfall in the region. On the basis of these factors the chief engineer decides, the length of the track to be overhauled but normally the plan is so drawn that the systematic overhauling of a section is completed in about 3 to 4 years. The stretch of track to be tackled in a particular year should be in continuation of the length overhauled during the previous year. If possible, gap adjustment, including joint survey and adjustment of creep, should be done prior to systematic overhauling.

## Lubrication of Rail Joints

The lubrication of rail joints is an important part of the work done on the permanent way and is incidental to systematic track maintenance. Joints are lubricated for the following purposes:

a) To allow for free expansion and contraction of rail.

b) To reduce wear and tear on the fishing planes of rails and the fish plates All rail joints, are lubricated once a year during the moderate season. This is also known as the oiling and greasing of fish plates. This work is not done during the rainy season. The lubricant used is a paste of workable consistency that consists of the following proportions of plumbago, kerosene oil, and black oil.

Plumbago (dry graphite) 5 kg Kerosene oil (second quality) 3.5 L Black or reclaimed oil 2.75 L.

For 100 joints of 52 kg/90 R or for 125 joints of 75 R/90 R sometimes only plumbago and kerosene oil are used in the ratio of 3:2 for lubricating fish plates. Black oil is, however, used for oiling fish bolts and nuts.

Rail joints should be lubricated only after ensuring that their surfaces are properly cleaned, preferably with the help of wire brushes and clean jute. Joints should not be lubricated in extreme temperatures or when the rails are in tension as a result of creep. Joints should be opened one at a time for lubrication. Even when opening a joint, only one fish plate should be tackled at a time and at no time during the operation should there be less than one fish plate and three fish bolts connecting the two rails.

## Picking up Slacks

Slacks are those points in the track where the running of trains is faulty or substandard. Slacks generally occur in the following cases:

a) Stretches of yielding formation.

b) Poorly maintained sections that have loose packing, bad alignment, and improper longitudinal and cross levels.

c) Improperly aligned curves.

d) Approaches to level crossings, girder bridges, etc., particularly in sags.

e) Portions of track with poor drainage.

f) Sections with an inadequate or unclean ballast cushion.

g) Other miscellaneous reasons.

In every working season, a certain number of days in each week (normally one or two days) are allotted to the picking up of slacks, depending upon the monsoon pattern and other local conditions. However, no through packing is done during the rainy season and slacks are only picked up in order to keep the track safe and in good running condition. In areas with less than 750 mm of rainfall, the alloted time may not be used only to attend to the slack but also to carry out through packing.

Slack may be sometimes picked up by packing only the following segments of the track:

a) Joint sleepers and the two other sleepers on either side of the joint, i.e., first shoulder and second shoulder sleepers.

b) A few sleepers in the approaches to level crossings and bridges.

c) Intermediate sleepers.

d) Stretches of track that adversely affect the running of trains as revealed by inspection notes.

It may be noted here that points and crossings should be attended to throughout the year. In sections with no points and crossings, the alloted time may be utilized for creep adjustment and other such track maintenance work. Two separate charts, one for main line work and another for yard work, are maintained by each gang and kept in the personal custody of the gang mate.

A recommended annual programme is drafted for regular track maintenance, in which each major activity is specified a certain time slot as per a fixed timetable (calendar). Table shows the timetable for regular track maintenance. Concrete sleeper tracks are maintained by on-track tie tamping machines.

Table: Timetable for regular track maintenance.

| Period | Work to be done |
|---|---|
| Post-monsoon: For about six months after the end of the monsoon. | Run down lengths in the entire gang beat to be dealt with in order to restore the quality of the section.<br><br>One cycle of systematic through packing/systematic directed track maintenance from one end of the gang length to the other, including overhauling of nominated sections.<br><br>Normally 4-5 days per week should be allotted for through packing/overhauling and the remaining days should be set aside for picking up slacks and for attending to bridge approaches, level crossings, and points and crossings over the entire gang beat. Other essential maintenance work such as lubrication of rail joints, joint gap adjustment, and realignment of curves should also be done during this period. |
| Pre-monsoon: For about two months prior to the break of the monsoon | Normally 2-4 days in a week should be devoted to clearing side and catch water drains, earthwork repairs of the cess, clearing waterways, and picking up slacks. Normal systematic maintenance should be carried out during the remaining days. |
| During the monsoon: For about four months. | Attention given to the track as required. This will consist primarily of picking up slacks and attention to side and catch water drains and waterways. During abnormally heavy rains, the line should be patrolled by the gangs in addition to regular monsoon patrolling. |

## Modern Methods of Track Maintenance

Modem methods of track maintenance employ track machines and other modem track equipment for the maintenance of tracks as opposed to the traditional methods of manual packing. The methods used generally are the following:

a) Mechanized maintenance of track with the help of track machines.

b) Measured shovel packing (MSP).

c) Directed track maintenance (DTM), which is maintenance based on need, particularly if DTM is done by MSP.

In these methods, the emphasis is not only on modern maintenance techniques but also on the identification and rectification of defects, the implementation of proper quality control in a systematic manner, and the use of proper tools and equipment.

For over a century, tracks on Railways have been maintained by the method of 'beater packing' using manual labour and the performance has been excellent. It is now being felt that this method of manual maintenance, which has stood the test of time, may require a revision in view of all the recent technological and social advancements. The European Railways, which was also in a similar position a few decades ago, has already switched over to modern methods of maintenance to suit the requirements of the modern track. Such a system of maintenance of track has given the men working on the permanent way a better social status, as it requires lesser physical strength and higher mechanical skill on their part. The system has also enabled them to maintain the modern track more economically and effectively in order to cater to the requirements of higher speeds and heavier axle loads. There is no doubt that the mechanized maintenance of track has become a technical necessity for the modern track structure with its long welded rails and concrete sleepers.

## Mechanized Methods of Track Maintenance

The need to switch over from manual to mechanized methods of track maintenance is progressively being felt due to the following reasons:

a) Beater packing is a very hard and strenuous job and thus the labourers have a tendency to shirk from doing this type of work.

b) It is difficult to ensure the uniform quality of the compaction under the sleepers carried out by manual means due to the uncertainties associated with the varying physical strength of the labourers, commitment of the workers, varying weather conditions, and other allied factors.

c) The intensity of the pressure and shock that the ballast is subjected to when the beater is being used is very high and in many cases exceeds the crushing strength of the stone. This results in the progressive clogging of the ballast section.

d) Traffic densities, axle loads, and speeds have increased considerably on Indian Railways considerably in the recent past. Beater packing does not enable track geometry to be maintained within the tolerances prescribed for a satisfactory length of time.

e) The retention of the packing done via manual maintenance is not very good and the track geometry gets distorted in a short time due to high-speed traffic.

f) Manual maintenance is not much suited to the modern track, which consists of long welded rails and heavy concrete sleepers.

g) With the increase in traffic density, the time available between trains is becoming progressively short. It is, therefore, becoming increasingly difficult to maintain tracks by manual methods, which take a considerably long time.

h) When a track is maintained manually, it takes considerable time for it to get fully consolidated and, therefore, speed restrictions exist for a long period after track renewal work has been completed.

Manual methods do not emphasize on the identification of defects and monitoring of the work being done. These are, however, done in the case of modern methods of track maintenance, thereby giving move effective results.

Table: Track machines in use on IR.

| Type of track machine | Functions | Remarks and output |
|---|---|---|
| Plassermatic 06-16-USLC | Tamping, levelling, and aligning. Consists of 16 tamping tools and can tamp one sleepers at a time. | Output is about 700 sleepers per hour. |
| Plassermatic 08 Unomatic tamper | Tamping, levelling, and lining simultaneously. Consists of 16 tamping tools and can tamp one sleeper at a time. | Output is 1000 sleepers per hour. |
| Plassermatic 08 Dnomatic tamper | Tamping, levelling, and lining simultaneously. Consists of 32 tamping tools and can tamp two sleepers at a time. | Output is 2000 sleepers per hour. |
| Plasser 09 CSM tamper | Tamping, levelling, and lining simultaneously. The machine advances continuously, while tamping is done from sleeper to sleeper. | Latest track machine with an output of 2500 sleepers per hour. |
| Plasser 08-275 unimat P and C tamper | Tamping points and crossings. Consists of 16 independent tiltable tamping tools, which can tamp points and crossings as well as plain tracks. | Good machine for tamping points and crossings. Can tamp 1 turnout in a 4-hour block. |
| Plasser VDM-800 consolidator | Used for crip and shoulder consolidating so as to improve the retention of packing. | Output is 1000 sleepers per hour. |
| DGS-62 N stabilizer | Dynamic track stabilizer that uses vibrating rollers to further stabilize the track. | Output is 500 sleepers per hour. |
| Matisa R-7 ballast regulator | To regulate ballast profile by transferring ballast from one place to another. | Can regulate ballast of 1 km of track per hour. |
| Plasser RM-80 BCM | Ballast cleaning machine used to excavate, clean, and put the screened ballast back in the track and to remove unwanted material. | Can handle 650 $m^3$ of ballast per hour. |
| Switch relaying machine Wm-22 | To remove and install parts or complete assemblies of points and crossings. | Can relay about two sets of points and crossings every day. |
| Track relaying trains (TRT) | Relaying the entire track automatically with very little manual effort. Consists of four different units and other ancillary machines with a total cost of over Rs 100 million. | Latest state-of-the art machine. Can accomplish the relaying of 1 km of track per day in a 4-hour block. |

On railways, the mechanized maintenance of tracks normally involves the use of mechanical tampers, which are used to tamp or pack the ballast. Mechanical tampers are generally of two types, namely, off-track tamper and on-track tampers.

## Necessity and Advantages of Track Maintenance

The railway track should be maintained properly in order to enable trains to run safely at the highest permissible speeds and to provide passengers a reasonable level of comfort during the ride. Track maintenance becomes a necessity due to following reasons:

a) Due to the constant movement of heavy and high-speed trains, the packing under the sleepers becomes loose and track geometry gets disturbed. The gauge, alignment, and longitudinal as well as cross levels of the track thus get affected adversely and the safety of the track is jeopardized.

b) Due to the vibrations and impact of high-speed trains, the fittings of the track come undone and there is heavy wear and tear of the track and its components.

c) The track and its components get worn out as a result of the weathering effect of rain, sun, and sand.

Track maintenance ensures that such situations do not arise. The other advantages of track maintenance are as follows:

a) If the track is suitably maintained, the life of the track as well as that of the rolling stock increases since there is lesser wear and tear of their components.

b) Regular track maintenance helps in reducing operating costs and fuel consumption.

c) Small maintenance jobs done at the appropriate time such as tightening a bolt or key, hammering the dog spike, etc., helps in avoiding loss of the concerned fitting and thus saving on the associated expenditure.

d) When track maintenance is neglected for a long time, it may render the track beyond repair, calling for heavy track renewals that entail huge expenses.

## Inspection and Maintenance of Points and Crossings

Points and crossings should be inspected in detail, as the quality of a train ride greatly depends on their maintenance. The following important points should be checked.

Condition of tongue rails and stock rails there should be no wear on the top as well as the gauge face side of the tongue rail. Badly worn out rails should be replaced. It should be ensured that the turnout side stock rail is provided with the requisite bend ahead of the toe of the switch; otherwise the alignment at this spot is bound to be kinky.

Condition of fittings of tongue and stock rails. The fittings should be tight and the spherical washers must be placed at their correct locations. The slide chairs should be cleaned and greased with graphite for smooth operation of the points. The fish plates should be provided with the correct amount of bend at the loose heel joint. A gauge tie plate should be added if provisions for the same have not been made.

Gauge and cross level at switch assembly. The gauge and cross levels should be checked for correctness at the following locations: (i) the stock joint, (ii) 150 mm (6") behind the toe of the switch, (iii) the mid-switch for the straight track and for the turnout side, and (iv) the heel of the switch for the straight track and for the turnout side.

Clearance between stock and tongue rails at the heel of the switch. The correct divergence to be provided at heel of the switch should be as follows:

1 in 16 or 1 in 12 1 in 8.5

BG-133 mm (5.25") 120 mm (4.25")

MG-117 mm (4.65") 120 mm (4.75")

Throw of the switch. The throw of the switch should be as follows:

Recommended Minimum

BG 115 mm (4.5") 95 mm (3.25")

MG 100 mm (4") 89 mm (3.5")

Condition of crossing and tongue rail. The condition of the crossings and of the fittings should be checked. The maximum vertical wear permitted on a point or wing rail is 10 mm and these should be reconditioned when the wear is 6 mm. The burn burrs should also be removed and the fittings should be tightened. The maximum vertical wear permitted on a tongue rail is 6 mm, whereas the permitted lateral wear is 8 mm for 90 R and 52-kg rails and 5 mm for 60 R and 75 R rails. The tongue rail should be replaced or reconditioned before this value is reached. The Railway Board has recently decided that the maximum vertical wear on wing rails and the nose of the crossings should be limited to 4 mm on the Rajdhani and Shatabdi routes and 6 mm on all other routes. The wear limits for CMS crossings are, however, 5.5 mm for Rajdhani and Shatabdi routes and 7.5 mm for all other routes.

Gauge and cross level of crossing assembly. The gauge and cross level should be checked at the following locations and should always be correct: (i) 1 m ahead of the nose on straight tracks and on turnouts, (ii) 150 mm (6") behind the ANC on straight tracks and on turnouts, and (iii) 1 m behind the ANC on straight tracks and on turnouts.

Check rails the condition of check rails should be ascertained. Check rail clearances should be as follows:

Maximum Minimum

BG 48 mm 44 mm

MG 44 mm 41 mm

Lead curvature the curvature should be checked either by the offset method or by the versine method. The curvature should be correct and uniform.

Cross levels on straight tracks and turnouts. The cross levels on straight tracks and turnouts should be checked to see that they are correct at all places.

Sleepers the condition of the sleepers and their fittings should be checked and unserviceable sleepers should be replaced. The squaring and spacing of sleepers should be proper and they should be well packed.

Ballast and drainage enough quantity of ballast should be available so as to provide an adequate cushion. The drainage should be proper.

Any other defects, if there are any other defects in the layout, these should be checked and corrected.

Special attention is required to ensure that the sleepers are well packed, all fittings are tightened, gauge and cross levels are properly maintained, and the wear on the tongue rail as well as on the crossing is within permissible limits. It should also be ensured that proper distance blocks are provided at correct locations. The schedule of inspection followed on Indian Railways is given in table.

| | |
|---|---|
| PWI (permanent way inspector) | Once in 3 months in rotation for passenger lines and once in 6 months in rotation for other lines. The interval between two. |
| III and PWI (in charge) | inspections for passenger lines of the same turnout should not exceed 4 months. |
| AEN (assistant engineer) | Once in 12 months for all passenger lines and test checking of 10% of other points and crossings. |
| DEN (divisional engineer) | Test checking of certain number of points and crossings, particularly in running lines and those recommended for renewals. |

## Maintenance of Rail Surface

The surface of rails is susceptible to certain defects in the absence of proper care and maintenance. Each defect should be dealt with carefully at the proper time.

### Hogged Joints

The hogging of a joint is a phenomenon in which the two rail ends at the joint get depressed on account of the poor maintenance of the rail joint, loose and faulty fastenings, and other such reasons. The hogging ofjoints brings about deterioration in the running quality of the track. The various techniques for removing this defect are as follows:

a) By measured shovel packing.

b) If hogging occurs on a small scale, the joint sleepers are overpacked and the fittings are tightened, which may finally remedy the defect. Use of liners and shims may also help in removing the defect in case the fittings are loose.

c) By cropping the hogged portions of the rail (preferably 450 cm), reusing the rail after the necessary drilling of holes, etc.

d) By dehogging the rail, i.e., removing the vertical bends on the reverse side of the rail with the help of a dehogging machine.

## High Joints

High joints result in a very uncomfortable ride on the track. High joints are the outcome of the following:

a) Changes in track structure, e.g., provision of wooden sleepers in a track that is normally laid with metal sleepers. Since it is easier to maintain wooden sleepers as compared to metal sleepers, this discrepancy in the type of sleepers results in high joints, which in turn produces the effect of camel back riding.

b) Sinking of intermediate sleepers.

c) Overpacking of joint sleepers.

This defect is removed by lifting and packing the intermediate sleepers.

## Blowing and Pumping Joints

A joint is called a blowing joint when it blows out fine dust during the passage of a train. The surroundings of such a joint are always coated with fine dust. A blowing joint becomes a pumping joint during the rainy season when it pumps out mud and water from the mud pockets formed below the joint. This defect is caused because of poor maintenance of the joint, particularly of the packing of the joint sleepers, unclean ballast, and bad drainage, and also sometimes due to surface defects in the rail such as scabbing. As moving loads pass over the joint, the joint sleepers get depressed and lifted up constantly. As this happens, the dust or mud gets sucked up and spreads in the vicinity of the joint. The remedy lies in:

a) Deep screening the ballast below the joint and shoulder sleepers.

b) Packing the joint sleeper and shoulder sleepers thoroughly.

c) Providing proper drainage at the joint.

d) Tightening loose fittings.

e) Adjusting the creep, if any.

## Lifting of Track

Normally, lifting a track becomes necessary when the track undergoes regrading. This may be due to yard remodelling, construction of a bridge, etc., or in an effort to eliminate the sags that develop in the approaches to level crossings and bridges and at other locations made vulnerable due to defective maintenance or yielding formation. The points regarding the lifting of tracks that require special mention are as follows:

a) A maximum of 75 mm (3 ") of the track should be lifted at a time. Whenever heavy lifting is involved, it should be done in different stages, with each lift being not more than 75 mm.

b) On single lines, lifting should commence from the downhill gradient and continue in the direction of the rising gradient. In the case of double lines, it should proceed in the direction opposite to that of the traffic, taking care not to exceed the easement grade.

c) Lifting should be done under the supervision of a PWI after imposing suitable speed restrictions and setting up the obligatory engineering signals.

## Lowering of Track

The lowering of a track becomes necessary when the track is re-graded for various reasons such as yard remodelling, provision of level crossings, etc. It should be avoided until it becomes inevitable, as lowering the track makes it unstable and is quite a difficult, time-consuming, and costly proposal.

## Longitudinal Sag in the Track

Normally a track between two rigid structures such as bridges, level crossings, etc., settles due to the passage of moving loads. The settlement of the track also takes place on yielding formation owing to the weakness of the formation and the puncturing of the ballast into the formation. Whereas a longitudinal unevenness in the shape of a vertical curve may not be noticeable, an irregular longitudinal sag may make the ride on the track uncomfortable. In such cases, a proper survey of the track should be carried out, pegs should be fixed at the correct longitudinal level with the help of a levelling instrument, and the track should then be lifted. The track should not be lifted more than 75 mm at a time. An adequate quantity of ballast should be collected in advance so that both packing and lifting can be done effectively. If excessive lifting is involved, the work should be done under speed restrictions.

## Centre-bound Sleeper

This defect is generally noticed in wooden and steel trough sleeper tracks. This defect occurs when, as a consequence of plying traffic, the sleeper starts to receive support at the centre instead of at the ends. If proper care in not taken during through packing and the middle portion of the sleeper is also packed, the defect can develop very early. Even under normal circumstances, the ballast under the sleeper ends, where the sleeper rests, gets more depressed compared to the ballast at the centre because of the impact of the moving loads and in the process the sleeper, instead of resting at the ends, stats to rest at the centre.

Centre binding of the sleepers leads to the rocking of the trains and is detrimental to the quality of the track. The defect can be removed by loosening the ballast at the centre of the sleeper. It is considered a good practice to make a small recess or depression in the ballast section at the centre of the sleeper.

## References


- Railroad-ref64406, technology: britannica.com, Retrieved 27 April, 2019
- “S-Bahn- und Fernverkehr Zürich–Olten: SBB saniert Fahrbahn im Heitersbergtunnel | SBB”. Www.sbb.ch (in German). 18 December 2013. Retrieved 2017-09-17
- Railroad-operations-and-control, railroad, technology: britannica.com, Retrieved 28 May, 2019
- Cole, George M; and Harbin; Andrew L (2006). Surveyor Reference Manual. Belmont, CA: Professional Publications Inc. P. 16. ISBN 1-59126-044-2

- Structure-of-the-railway-track-fig1-31624125, figure: researchgate.net, Retrieved 28 May, 2019
- "Tata Steel opens French plant to heat treat 108-meter train rail". International Organization on Shape Memory and Superelastic Technologies (SMST). ASM International. 30 October 2014. Retrieved 10 September 2014
- Requirement-of-track-material, railway-engineering: eng.dieselloc.ru, Retrieved 29 June, 2019
- Mundrey, J.S. (2000). Railway track engineering (3rd ed.). New Delhi: Tata mcgraw-Hill Pub. P. 165. ISBN 978-0-07-463724-1. Retrieved 14 November 2012
- Railway-engineering/specifications-for-track-ballast: eng.dieselloc.ru, Retrieved 31 August, 2019
- Bonnett, Clifford F. (2005). Practical Railway Engineering (2nd ed.). London, UK: Imperial College Press. ISBN 978-1-86094-515-1. OCLC 443641662
- What-is-ballast-railway-ballast-functions-of-ballast: engineeringarticles.org, Retrieved 1 January, 2019
- Federal Railroad Administration Track Safety Standards Compliance Manual (PDF). Federal Railroad Administration. 1 April 2007. Archived from the original (PDF) on 28 May 2008. Retrieved 13 November 2012

# 3
# Rail Gauge

The distance from the inside of one rail on a railroad track to the inside of the other is known as the rail gauge. The gauges are broadly classified on the basis of their size into standard-gauge, broad-gauge, Iberian-gauge, narrow-gauge and minimum-gauge. This chapter has been carefully written to provide an easy understanding of these types of rail gauges as well as their advantages and disadvantages.

The gauge of a railway track is defined as the clear minimum perpendicular distance between the inner faces of the two rails. Following are the factors affecting the choice of a gauge:

1. Traffic Condition: If the intensity of traffic on the track is likely to be more, a gauge wider than the standard gauge is suitable.
2. Development of Poor Areas: The narrow gauges are laid in certain parts of the world to develop a poor area and thus link the poor area with the outside developed world.
3. Cost of Track:
    - The cost of railway track is directly proportional to the width of its gauge.
    - If the fund available is not sufficient to construct a standard gauge, a metre gauge or a narrow gauge is preferred rather than to have no railways at all.
4. Speed of Movement:
    - The speed of a train is a function of the diameter of wheel which in turn is limited by the gauge.
    - The wheel diameter is usually about 0.75 times the gauge width and thus, the speed of a train is almost proportional to the gauge.
    - If higher speeds are to be attained, the broad gauge track is preferred to the metre gauge or narrow gauge track.
5. Nature of Country:
    - In mountainous country, it is advisable to have a narrow gauge of the track since it is more flexible and can be laid to a smaller radius on the curves.
    - This is the main reason why some important railways, covering thousands of kilometers, are laid with a gauge as narrow as 610 mm.

### Advantages and Disadvantages of Different Track Gauges

Narrow gauge railways usually cost less to build because they are usually lighter in construction, using smaller cars and locomotives (smaller loading gauge), as well as smaller bridges, smaller tunnels (smaller structure gauge) and tighter curves. Narrow gauge is thus often used in mountainous terrain, where the savings in civil engineering work can be substantial. It is also used in sparsely populated areas, with low potential demand, and for temporary railways that will be removed after short-term use, such as for construction, the logging industry, the mining industry, or large-scale construction projects, especially in confined spaces.

Broader gauge railways are generally more expensive to build, but are able to handle heavier and faster traffic.

750 mm narrow gauge railway in Waldenburg.

Meter gauge tracks, Istanbul nostalgic tramways.

In approximately 55% of the world's railways are used standard gauge (1435 mm).

Inclined lift with a gauge of 8.2 metres.

## Break of Gauge-Rail Gauges

The rail gauges selected to build a railroad is chosen for a variety of reasons. These may include a need to connect to other railroads, the type of terrain to be crossed and the costs of construction and maintenance.A railroad may start as a single isolated line, for example from a mine or quarry to a port, and be built using a gauge that is ideal for the purpose of the line and the conditions along its route. Later a mainline railroad, built using a different gauge chosen for equally valid reasons, connects the port to other cities and industrial centres. Where the two railroads with their different gauges meet is a "break of gauge" location.

Break of gauge locations are a source of complaints from passengers and an added expense to industries. While passengers can be easily moved, often with complaints, from their seats on one train to seats on another train, this is not always easily done for freight, particularly bulk freight. Losses and damage are common problems. On some journeys passengers and freight would change gauges several times. Schemes devised to overcome this problem have led to some very innovative, and sometimes strange, solutions.

The Penrhyn Quarry Railway, a 1 ft 10 ¾ in (578mm) gauge railway in Wales constructed for hauling slate.

## Separation of Gauges

The simplest method of dealing with break of gauge problems, and the method most commonly used in the early days of railroading, was to keep the two gauges completely separate. So each gauge would have its own platforms, loading docks, servicing facilities, etc. It was not uncommon for passengers to have to use a taxi to get from a station on one gauge to the station on another gauge to continue their journey. Likewise, freight would have to be transported between the different loading facilities by road wagons.

The break of gauge platform.

The image above shows the situation at Albury on the border between the states of New South Wales and Victoria in Australia. Railroads in Victoria were built using the Irish Broad Gauge (5 ft 3 in, 1600mm) while those in New South Wales were built using the Stephenson or Standard Gauge (4 ft 8 ½ in, 1435mm). Passengers moving from one system to the other simply crossed the platform to the train waiting on the other side to continue their journey. A similar but very labour intensive arrangement existed for transferring freight between the two systems.

## Multi Gauge Track

More commonly tracks of a different gauge or gauges would be laid within track of another gauge. This allowed trains of different gauges to share the same platforms, loading docks and servicing facilities but obviously made track switching and signaling more complicated and expensive.

The image above shows dual gauge track (in the foreground and extending into the distance) merging with triple gauge track coming from the right.The dual gauges in the foreground are Irish Broad Gauge and Cape Narrow Gauge (3 ft 6 in, 1067mm) which both share a common right hand rail.The triple gauge tracks merging from the right are Irish Broad Gauge (the first and third rails from the left), Cape Narrow Gauge (the second rail and the shared third rail from the left) and Standard Gauge (the second rail, which is shared with the narrow gauge, and the fourth rail).

Dual and triple gauge tracks.

A triple gauge turntable (Cape Narrow, Standard and Irish Broad) at Steam Town.

Using shared rails reduces the construction costs but is not always possible. If the difference between the gauges is small then there will not be enough room between the two outer (unshared) rails for the switches, rail plates and spikes. In the image of the turntable above the rails had to be custom made to fit the three gauges together without a common rail. Note the very small gap between the Irish Broad gauge (the outer rails) and the Standard gauge tracks.

Long distance dual gauge track in Lithuania without a shared rail. The first and third rails (from the left) are standard gauge. The second and forth rails are Russian broad gauge (4 ft 11 5/6 in, 1520mm).

One solution to the break of gauge problem is to replace a single gauge line, in its entire length, with dual gauge track. The image above shows one example where the gauges are too close to use a common rail. This allows passengers and freight to make the entire journey without having to change trains.

Break of gauge locations with four or more gauges are rare due to their costs and engineering difficulties. However, examples can be found in industrial systems, such as locomotive or rolling stock manufacturing plants and in railroad museums.

## Bogie Exchange

In a bogie exchange, the wagons are lifted off their current gauge bogies which are rolled away. The new gauge bogies are then rolled into place and the wagons lowered onto them. Most rolling stock (wagons and carriages) are designed to have their bogies replaced anyway so this system was not specifically designed for the gauge conversion of trains but it is ideal for the task. While the process can be time consuming the freight carried in the wagons does not have to be moved between wagons and this reduces losses and damage.

Bogie gauge exchange on the border between China and Russia.

Diesel locomotives can also be gauge converted by this method but their more complicated bogie connections makes the process more difficult and time consuming. Some steam locomotives built after 1930 were designed to have their bogies and driving wheels gauge converted but there are records of only one such conversion ever taking place – a Pacific class steam locomotive R766 in Victoria Australia, which was converted from Irish Broad gauge to Standard gauge in 2014.

## Piggyback" or "Train on Train"

Rail wagons are commonly used to transport cars, trucks and other road vehicles between destinations. In the Channel Tunnel, for example, road vehicles are driven onto specially designed wagons at one end of the tunnel, the train takes the vehicles through the tunnel and they are driven off the wagons onto roads at the other end. Exactly the same principle, with variations, has been used to move rail wagons and even entire trains of one gauge over territory where a different rail gauge is in use.

A standard gauge wagon on a narrow gauge transporter wagon.

The image above shows an example of a single wagon transport system designed to transport a wagon of one rail gauge over tracks of a different rail gauge. These systems are typically used in wagon repair facilities. In this case the upper wagons wheels are locked in position on rails that form part of the lower wagon.

A standard gauge wagon supported on a 750mm gauge bogie.

The "piggy back" method has, on rare occasions, been used to carry entire trains across a different gauge. In the late 1950s when the narrow gauge Maree line in South Australia was being gauge converted, a set of standard gauge flat wagons were fitted with narrow gauge tracks on their tops. Narrow gauge wagons loaded with coal were pushed up a ramp at the end of the standard gauge train and along the "piggy back" track on the flat wagons. The coal wagons were then transported to the other end of the gauge converted section where the process was reversed. This continued for the year it took to complete the conversion of the entire line to standard gauge and ensured that a major power station continued to receive its coal supply.

## Variable Gauge

There are a variety of systems in use that allow the wheels of a train to be changed from one rail gauge to another without using bogie exchange. The idea goes back to the 1800s when railroads in Great Britain were plagued with change of gauge problems. In the most advanced of these systems the entire train, usually without the locomotive unless it also has variable gauge wheels, rolls slowly through a gauge changing mechanism which unlocks the wheels, moves them closer (or further apart) on their axles and then locks them again in their new position.

The automated gauge conversion Talgo system at Lleida, Spain on the border with France.

In the image above, the train enters the shed at the far end on the Spanish side as an Iberian Gauge train (5 ft 5 21/32 in, 1668mm) and emerges at this end on the French side as a standard gauge train. Spain is the largest user of variable gauge systems.

# Variable Gauge System

A variable gauge system allows railway vehicles in a train to travel across a break of gauge caused by two railway networks with differing track gauges.

For through-operation, a train must be equipped with special trucks holding variable gauge wheelsets containing a variable gauge axle (VGA). The gauge is altered by driving the train through a gauge changer or gauge changing facility.

As the train passes through the gauge changer, the wheels are unlocked, moved closer together, or further apart, and are then re-locked. Installed variable gauge systems exist within the internal network of Spain, and are installed on international links between Spain/France (Spanish train), Sweden/Finland (Swedish train), Poland/Lithuania (Polish train) and Poland/Ukraine (Polish train).

A system for changing gauge, without need for stopping is widespread for passenger traffic in Spain, used in services run on a mix of dedicated high-speed lines (using Standard gauge) and older lines (using Iberian gauge). Similar systems for freight traffic are still rather incipient, as the higher axle weight increases the technological challenge. Although several alternatives exist, including transferring freight, replacing individual wheels and axles, truck exchange, transporter flatcars or the simple transshipment of freight or passengers, they are impractical, thus a cheap and fast system for changing gauge would be beneficial for cross-border freight traffic.

Alternative names include Gauge Adjustable Wheelsets (GAW), Automatic Track Gauge Change-over System (ATGCS/TGCS), Rolling Stock Re-Gauging System (RSRS), Rail Gauge Adjustment System (RGAS), Shifting wheelset, Variable Gauge Rolling Truck, track gauge change and track change wheelset.

Variable gauge axles help solve the problem of a break-of-gauge without having to resort to dual gauge tracks or transshipment. Systems allow the adjustment between two gauges. No gauge changer designs supporting more than two gauges are used.

## Systems

Variable gauge axle DR III for 1,435 mm (4 ft 8 $\frac{1}{2}$ in) and 1,524 mm (5 ft) gauge, developed in 1957.

There are several variable gauge axle systems:

- Talgo-RD (from Talgo).
    - The Talgo system has been in revenue service in Portbou and Irun, on the Spanish-French border, since 1968.
    - It is used on the Strizh train between Moscow and Berlin.

- CAF-BRAVA (from Construcciones y Auxiliar de Ferrocarriles).
    - The BRAVA system was originally designed in 1968 by the Vevey Company (since absorbed by Bombardier Transportation), a company located in the city of Vevey on Lake Geneva, in Switzerland. The system was originally called the "Vevey axle". The design was subsequently obtained and improved by Construcciones y Auxiliar de Ferrocarriles (CAF).
- DB Cargo–Knorr-Bremse. being developed in 2002 for use between Europe and Russia.
- DBAG–Rafil Type V for freight (from *Radsatzfabrik Ilsenburg* for Deutsche Bahn).
- Japan Railways RTRI (from the Japan Railway Technical Research Institute) to be used on motorised axles.
- PKP SUW 2000 system produced by ZNTK Poznań for Polish State Railways.
- The Montreux-Oberland Bernois (MOB) railway, Switzerland in preparation by Prose (Manufacturer) of Winterthur 1,435 mm (4 ft 8 $^{1}/_{2}$ in)/1,000 mm (3 ft 3 $^{3}/_{8}$ in).

## Compatibility

The variable gauge systems are not themselves all compatible. Only the SUW 2000 and Rafil Type V systems are interoperable.

In 2009, at Roda de Barà near Tarragona, a Unichanger capable of handling four different VGA systems was under development.

## International Traffic

VGA is particularly important with international railway traffic because gauge changes tend to occur more often at international borders.

## Speed

The maximum speed of the trains equipped with the different technologies varies. Only CAF and Talgo produce high-speed VGA, allowing speeds up to 250 km/h.

## Gauge Changer

A *Talgo* gauge changing system in Lleida.

A gauge changer is a device which forces the gauge adjustment in the wheels. Designs consist of a pair of running rails that gradually vary in width between the two gauges, combined with other rails and levers to unlock, move, support and re-lock the adjustable axles.

In the Spanish Talgo-RD system, a constant spray of water is used to lubricate the metal surfaces, to reduce heat and wear. A Talgo-RD gauge changer is 20 metres long and 6 metres wide.

## Operation

Variable gauge multiple units, or a train including a variable gauge locomotive (e.g. Talgo 250) and rolling stock, may drive straight across a gauge changer. Normally the locomotive will not be able to change gauge, meaning that it must move out of the way whilst the remainder of the train itself passes through. On the opposite side, a new locomotive of the other gauge will couple to the train.

A train (or an individual car) can be pushed halfway across the gauge-changer, uncoupled, and then (once far enough across) coupled to the new locomotive and pulled the rest of the way. A long length of wire-rope with hooks on the end means that the process can be asynchronous, with the rope used to bridge across the length of the gauge changer (to temporarily couple the arriving cars and receiving locomotive, although without braking control from the locomotive to the train vehicles).

On long-distance trains in Spain and night trains crossing from Spain into France, the arriving locomotive stops just short of the gauge changer, uncouples and moves into a short siding out of the way. Gravity then moves the train through the gauge changer at a controlled low speed. The new locomotive is coupled onto the front only after the full train has finished passing through the changer.

## Comparison with Truck Exchange

### Time Taken

In VGA, the train is pulled through the "adjuster" at about 10 km/h (2.77 m/s) without any need to uncouple the wagons or disconnect (and test) the brake equipment.

### Spares

VGA always has the exact number of wheels of each gauge and they are always at hand, whereas with truck exchange, there must be a complete set of trucks of the right gauge and with wheels of the same diameter at the depot.

Truck exchange needs a stock of spare trucks of each gauge, which need to be shunted as required. If there is an influx of traffic one might run out of trucks of the other gauge. Depending on the sophistication of the exchange facility, the cars may need to be uncoupled for the truck exchange to take place.

### Locomotives

Steam locomotive are generally not gauge convertible on-the-fly. While diesel locomotives can be truck exchanged, this is not normally done owing to the complexity in the reconnection of cables and hoses. In Australia, some locomotives are transferred between gauges. The transfer might happen every few months, but not for an individual trip.

By 2004, variable gauge electric passenger locomotives were available from Talgo. It is not clear if variable gauge freight locomotives are available.

### Electric

- L-9202 is an experimental high speed Bo-Bo dual voltage (3 kV DC/25 kV AC) VGA locomotive.
- Talgo 250 locomotives were also planned to haul dual-voltage variable-gauge trainsets from Montpellier 200 km from the border to Barcelona and Madrid. Two Talgo 250 power cars haul 11 passenger trailer cars.
- EMU.

### Weight

- A gauge adjustable truck complete with wheelsets weighs a total of about one ton/tonne more than a conventional truck and normally must use disc brakes, which cool more slowly.

## Track Gauges by Size

### Standard-gauge Railway

A standard-gauge railway is a railway with a track gauge of 1,435 mm (4 ft 8 ½ in). The standard gauge is also called Stephenson gauge after George Stephenson, International gauge, UIC gauge, uniform gauge, normal gauge and European gauge in the European Union and Russia. It is the most widely used railway track gauge across the world, with approximately 55% of the lines in the world using it. All high-speed rail lines use standard gauge except those in Russia, Finland, Portugal and Uzbekistan. The distance between the inside edges of the rails is defined to be 1435 mm except in the United States, where it is still defined in U.S. customary units as exactly "four feet eight and one half inches" (0.1 mm larger than the metric standard).

As railways developed and expanded, one of the key issues was the track gauge (the distance, or width, between the inner sides of the rails) to be used. Different railways used different gauges, and where rails of different gauge met – a "gauge break" – loads had to be unloaded from one set of rail cars and re-loaded onto another, a time-consuming and expensive process. The result was the adoption throughout a large part of the world of a "standard gauge" of 1435 mm (4 ft 8 ½ in), allowing interconnectivity and interoperability.

A popular legend that has been around since at least 1937 traces the origin of the 1435 mm (4 ft 8 ½ in) gauge even further back than the coalfields of northern England, pointing to the evidence of rutted roads marked by chariot wheels dating from the Roman Empire. It is curious that the Roman pace or *passus* was 4.855 ft or 1435 mm; a thousand such was one Roman mile. Snopes categorised this legend as "false", but commented that "it is perhaps more fairly labelled as 'True, but for trivial and unremarkable reasons'". The historical tendency to place the wheels of horse-drawn vehicles approximately 5 ft (1,524 mm) apart probably derives from the width needed to fit a carthorse in between the shafts. In addition, while road-travelling vehicles are typically measured from the outermost portions of the wheel rims (and there is some evidence that the first railways were measured in this way as well), it became apparent that for vehicles travelling on rails it was better to have the wheel flanges located *inside* the rails, and thus the distance measured on the inside of the wheels (and, by extension, the inside faces of the rail heads) was the important one.

There was never a standard gauge for horse railways, but there were rough groupings: in the north of England none was less than 4 ft (1,219 mm). Wylam colliery's system, built before 1763, was 5 ft (1,524 mm), as was John Blenkinsop's Middleton Railway; the old 4 ft (1,219 mm) plateway was relaid to 5 ft (1,524 mm) so that Blenkinsop's engine could be used. Others were 4 ft 4 in (1,321 mm) (in Beamish) or 4 ft 7 $^1/_2$ in (1,410 mm) (in Bigges Main (in Wallsend), Kenton, and Coxlodge).

The English railway pioneer George Stephenson spent much of his early engineering career working for the coal mines of County Durham. He favoured 4 ft 8 in (1,422 mm) for wagonways in Northumberland and Durham, and used it on his Killingworth line. The Hetton and Springwell wagonways also used this gauge.

Stephenson's Stockton and Darlington railway (S&DR) was built primarily to transport coal from mines near Shildon to the port at Stockton-on-Tees. The initial gauge of 4 ft 8 in (1,422 mm) was set to accommodate the existing gauge of hundreds of horse-drawn chaldron wagons that were already in use on the wagonways in the mines. The railway used this gauge for 15 years before a change was made to the 4 ft 8 $^1/_2$ in gauge. The historic Mount Washington Cog Railway, the world's first mountain-climbing rack railway, is still in operation in the 21st century, and has used the earlier 4 ft 8 in gauge since its inauguration in 1868.

George Stephenson used the 4 ft 8 $^1/_2$ in gauge (including a belated extra $^1/_2$ in (12.7 mm) of free movement to reduce binding on curves) for the Liverpool and Manchester Railway, authorised in 1826 and opened 30 September 1830. The success of this project led to Stephenson and his son Robert being employed to engineer several other larger railway projects. Thus the 4 ft 8 $^1/_2$ gauge became widespread and dominant in Britain. Robert was reported to have said that if he had had a second chance to choose a standard gauge, he would have chosen one wider than 4 ft 8 $^1/_2$ in (1,435 mm). "I would take a few inches more, but a very few".

During the "gauge war" with the Great Western Railway, standard gauge was called narrow gauge, in contrast to the Great Western's 7 ft $^1/_4$ in (2,140 mm) broad gauge. The modern use of the term "narrow gauge" for gauges less than standard did not arise for many years, until the first such locomotive-hauled passenger railway, the Ffestiniog Railway was built.

## Adoption

In 1845, in the United Kingdom of Great Britain and Ireland, a Royal Commission on Railway Gauges reported in favour of a standard gauge. The subsequent Gauge Act ruled that new passenger-carrying railways in Great Britain should be built to a standard gauge of 4 ft 8 $^1/_2$ in (1,435 mm), and those in Ireland to a new standard gauge of 5 ft 3 in (1,600 mm). In Great Britain, Stephenson›s gauge was chosen on the grounds that existing lines of this gauge were eight times longer than those of the rival 7 ft (2,134 mm) (later 7 ft $\frac{1}{4}$ in (2,140 mm)) gauge adopted principally by the Great Western Railway. It allowed the broad-gauge companies in Great Britain to continue with their tracks and expand their networks within the "Limits of Deviation" and the exceptions defined in the Act. After an intervening period of mixed-gauge operation (tracks were laid with three rails), the Great Western Railway finally completed the conversion of its network to standard gauge in 1892. In North East England, some early lines in colliery (coal mining) areas were 4 ft 8 in (1,422 mm), while in Scotland some early lines were 4 ft 6 in (1,372 mm). All these lines had been widened to standard gauge by 1846. The British gauges converged starting from 1846 as the advantages of equipment interchange became increasingly apparent. By the 1890s, the entire network was converted to standard gauge.

The Royal Commission made no comment about small lines narrower than standard gauge (to be called "narrow gauge"), such as the Ffestiniog Railway. Thus it permitted a future multiplicity of narrow gauges in the UK. It also made no comments about future gauges in British colonies, which allowed various gauges to be adopted across the colonies.

Parts of the United States, mainly in the Northeast, adopted the same gauge, because some early trains were purchased from Britain. The American gauges converged, as the advantages of equipment interchange became increasingly apparent. Notably, all the 5 ft (1,524 mm) broad gauge track in the South was converted to "almost standard" gauge 4 ft 9 in (1,448 mm) over the course of two days beginning on 31 May 1886.

In continental Europe, France and Belgium adopted a 1,500 mm (59 in) gauge from axis to axis of rail for their early railways. The gauge between the interior edges of the rails (the measurement adopted from 1844) differed slightly between countries, and even between networks within a country (for example, 1440 to 1445 mm in France).

The first tracks in Austria and in the Netherlands had other gauges (1,000 mm or 3 ft 3 $\frac{3}{8}$ in in Austria for the Donau Moldau linen and 1,945 mm or 6 ft 4 $\frac{9}{16}$ in in the Netherlands for the Hollandsche IJzeren Spoorweg-Maatschappij), but for interoperability reasons (the first rail service between Paris and Berlin began in 1849, first Chaix timetable) Germany adopted standard gauges, as did most other European countries.

The modern method of measuring rail gauge was agreed in the first Berne rail convention of 1886, according to the "Revue générale des chemins de fer, July 1928".

## Early Railways by Gauge

### Non-standard Gauge

- Monkland and Kirkintilloch Railway, authorised 1824 and opened 1825, used 4 ft 6 in (1,372 mm).
- Dundee and Newtyle Railway, authorised 1829 and opened 1831, used 4 ft 6 $\frac{1}{2}$ in (1,384 mm).
- The Eastern Counties Railway, authorised on 4 July 1836, used 5 ft (1,524 mm).

- The London and Blackwall Railway, authorised on 28 July 1836, used 5 ft ½ in (1,537 mm).
- The Dundee and Arbroath Railway, incorporated on 19 May 1836 and opened October 1838, used 5 ft 6 in (1,676 mm) until standardised in 1847.
- The Arbroath and Forfar Railway, incorporated on 19 May 1836 and opened November 1838, used 5 ft 6 in (1,676 mm).
- The Northern and Eastern Railway, authorised on 4 July 1836, used 5 ft (1,524 mm) gauge.
- Aberdeen Railway, opened 1848, used 5 ft 6 in (1,676 mm) until standardised.

## Almost Standard Gauge

- The Killingworth colliery railway, used 4 ft 8 in (1,422 mm).
- The Hetton colliery railway, opened 1822, used 4 ft 8 in (1,422 mm).
- The Stockton and Darlington Railway, authorised 1821, opened 1825, used 4 ft 8 in (1,422 mm).
- The Manchester and Leeds Railway, authorised on 4 July 1836, used 4 ft 9 in (1,448 mm).
    - The 4 ft 9 in (1,448 mm) railways were intended to take 4 ft 8 ½ in (1,435 mm) gauge vehicles and allow a (second) running tolerance.
- The Chester and Birkenhead Railway, authorised on 12 July 1837, used 4 ft 9 in (1,448 mm).
- The London and Brighton Railway, authorised on 15 July 1837, used 4 ft 9 in (1,448 mm).
- The Grand Junction Railway, authorised 1833, opened 1837, connected to LMR.
- The Manchester and Birmingham Railway, authorised on 30 June 1837, used 4 ft 9 in (1,448 mm).
- The trams in Dresden, authorised in 1872 as horsecars, used 1,440 mm (4 ft 8 $^{11}/_{16}$ in) gauge vehicles. Converted to 600 V DC electric trams in 1893, they now use 1,450 mm (4 ft 9 $^{3}/_{32}$ in); both gauges are within the tolerance for standard gauge.
- The trams in Nuremberg nominally used 1,432 mm (4 ft 8 ⅜ in) during much of their existence, but have since been converted to standard gauge in name as well as fact.

## Standard Gauge

- The Liverpool and Manchester Railway, authorised 1824, opened 1830.
- The Saint-Étienne–Lyon railway, authorised 1826, opened 1833 ( all the early French railways including Saint-Etienne Andrezieux, authorised 1823, opened 1827 had a French Gauge of 1,500 mm (4 ft 11 $^{1}/_{16}$ in) from rail axis to rail axis, compatible with early standard gauge tolerances).
- The Newcastle & Carlisle Railway, authorised 1829, opened 1834, isolated from LMR.

- The Grand Junction Railway, authorised 1833, opened 1837, connected to LMR.
- The London and Birmingham Railway, authorised 1833, opened 1838, connected to LMR.
- The Manchester and Birmingham Railway, authorised 1837, opened 1840, connected to LMR.
- The Birmingham and Gloucester Railway, authorised 1836, opened 1840, connected to LMR.
- The London and Southampton Railway, authorised 1834, opened 1840.
- The London and Brighton Railway, authorised 1837, opened 1841.
- The South Eastern Railway, authorised 1836, opened 1844.

### Dual Gauge

- Cheltenham and Great Western Union Railway, authorised 1836, opened 1840, dual gauge 1843 4 ft 8 ½ in (1,435 mm) standard gauge and 7 ft ¼ in (2,140 mm).

### Modern almost Standard Gauge Railways

- The Toronto Transit Commission uses 1,495 mm (4 ft 10 ⅞ in) gauge on its streetcar and subway lines, which was actually closer to 1,520 mm (4 ft 11 27/32 in) gauge. However, the Eglinton Crosstown Line will use 1,435 mm (4 ft 8 ½ in) gauge.
- Trams in Dresden, Germany use 1,450 mm (4 ft 9 3/32 in).
- Trams in Leipzig, Germany use 1,458 mm (4 ft 9 13/32 in).
- 1,445 mm (4 ft 8 ⅞ in) gauge is in use on several urban rail transit systems in Europe:
  - Trams in Italy.
  - Madrid Metro (only metro system. Light rail system uses 1,435 mm (4 ft 8 ½ in) gauge).
- The MTR in Hong Kong uses 1,432 mm (4 ft 8 ⅜ in) gauge on lines owned by the MTR Corporation. However, lines formerly operated (but which continue to be owned) by the Kowloon-Canton Railway Corporation, including the Light Rail network, use 1,435 mm (4 ft 8 ½ in) gauge.
- The Bucharest Metro uses 1,432 mm (4 ft 8 ⅜ in) gauge.
- The Washington Metro uses 4 ft 8 ¼ (1,429 mm), ¼ in (6 mm) narrower than standard gauge.
- The Mount Washington Cog Railway, the world's oldest mountain-climbing rack-and-pinion railway, uses a 4 ft 8 in (1,422 mm) gauge.

### Broad-gauge Railway

A broad-gauge railway is a railway with a track gauge broader than the 1,435 mm (4 ft 8 ½ in) standard-gauge railways.

Great Western Railway broad-gauge steam locomotives awaiting scrapping in 1892 after the conversion of the tracks to 4 ft 8 ½ in (1,435 mm) standard gauge.

A replica GWR Firefly Class broad gauge steam locomotive, with a preserved, standard-gauge GWR railcar, Didcot Railway Centre.

Broad gauge was first used in Great Britain in Scotland for two short, isolated lines, the Dundee and Arbroath Railway (1836-1847) and the Arbroath and Forfar Railway (1838-1848). Both the lines were built in 5 ft 6 in (1,676 mm). Both the lines were subsequently converted to standard gauge and connected to the emerging Scottish rail network.

The Great Western Railway, was designed by Isambard Kingdom Brunel, in 1838, with a gauge of 7 ft ¼ in (2,140 mm), and retained this gauge until 1892. Some harbours also used railways of this gauge for construction and maintenance. These included Portland Harbour and Holyhead Breakwater, which used a locomotive for working sidings. As it was not connected to the national network, this broad-gauge operation continued until the locomotive wore out in 1913. The gauge initially proposed by Brunel was 7 ft (2,134 mm) exactly but this was soon increased by ¼ in (6 mm) to 7 ft ¼ in (2,140 mm) to accommodate clearance problems identified during early testing. George Stephenson was to add an extra half inch to his original 4 ft 8 in (1,422 mm) gauge for the same reason.

While the parliament of the United Kingdom of Great Britain and Ireland was initially prepared to authorise lines built to the broad gauge of 7 ft (2,134 mm), it was eventually rejected by the Gauge Commission in favour of all new railways in England, Wales and Scotland being built to standard gauge of 4 ft 8 ½ in (1,435 mm), this being the gauge with the greatest mileage. Railways which had already received their enabling Act would continue at the 7 ft gauge. Ireland, using the same criteria, was allocated a different standard gauge, the Irish gauge, of 5 ft 3 in (1,600 mm) which is also used in the Australian states of South Australia and Victoria. Broad-gauge lines in Britain were gradually converted to dual gauge or standard gauge from 1864, and finally the last of Brunel's broad gauge was converted over a single weekend in 1892.

In 1839 the Netherlands started its railway system with two broad-gauge railways. The chosen gauge of 1,945 mm (6 ft 4 $\frac{9}{16}$ in) was applied between 1839 and 1866 by the Hollandsche IJzeren Spoorweg-Maatschappij (HSM) for its Amsterdam–The Hague–Rotterdam line and between 1842 and 1855, firstly by the Dutch state, but soon by the Nederlandsche Rhijnspoorweg-Maatschappij, for its Amsterdam–Utrecht–Arnhem line. But the neighbouring countries Prussia and Belgium already used standard gauge, so the two companies had to regauge their first lines. In 1855, NRS regauged its line and shortly afterwards connected to the Prussian railways. The HSM followed in 1866. There are replicas of one broad-gauge 2-2-2 locomotive (*De Arend*) and three carriages in the Dutch Railway Museum in Utrecht. These replicas were built for the 100th anniversary of the Dutch Railways in 1938–39.

Ireland and some states in Australia and Brazil have a gauge of 5 ft 3 in (1,600 mm), but Luas, the Dublin light rail system, is built to standard gauge. Russia and the other former Soviet Republics use a 1,520 mm (4 ft 11 $\frac{27}{32}$ in) (originally 5 ft (1,524 mm)) gauge while Finland continues to use the 5 ft (1,524 mm) gauge inherited from Imperial Russia (the two standards are close enough to allow full interoperability between Finland and Russia). Portugal and the Spanish *Renfe* system use a gauge of 1,668 mm (5 ft 5 $\frac{21}{32}$ in) called *Ancho Ibérico* in Spanish or *Bitola Ibérica* in Portuguese though there are plans to convert to standard gauge.

In Toronto, Canada, the gauge for TTC subways and streetcars was chosen in 1861. Toronto adopted a unique gauge of 4 ft 10 $\frac{7}{8}$ in (1,495 mm), an "overgauge" originally stated to "allow horse-drawn wagons to use the rails", but with the practical effect of precluding the use of standard-gauge equipment in the street. The Toronto Transit Commission still operates the Toronto streetcar system and three subway lines on its own unique gauge of 4 ft 10 $\frac{7}{8}$ in (1,495 mm). The Scarborough RT, however, uses standard gauge, as will the future light rail lines of the Transit City plan (including the under-construction Eglinton Crosstown line).

In 1851 the 5 ft 6 in (1,676 mm) broad gauge was officially adopted as the standard gauge for the Province of Canada, becoming known as the *Provincial gauge*, and government subsidies were unavailable for railways that chose other gauges. This caused problems in interchanging freight cars with northern United States railroads, most of which were built to standard gauge or a gauge similar to it. In the 1870s, mainly between 1872 and 1874, Canadian broad-gauge lines were changed to standard gauge to facilitate interchange and the exchange of rolling stock with American railroads. Today, all Canadian railways are standard-gauge.

In the early days of rail transport in the US, railways tended to be built out from coastal cities into the hinterland, and systems did not initially connect. Each builder was free to choose its own gauge, although the availability of British-built locomotives encouraged some railways to be built to standard gauge. As a general rule, southern railways were built to one or another broad gauge, mostly 5 ft (1,524 mm), while northern railroads that were not standard gauge tended to be narrow gauge. Most of the original track in Ohio was built in 4 ft 10 in (1,473 mm) Ohio gauge, and special "compromise cars" were able to run on both this track and standard gauge track. In 1848, Ohio passed a law stating "The width of the track or gauge of all roads under this act, shall be four feet ten inches between the rails". When American railroads track extended to the point that they began to interconnect, it became clear that a single nationwide gauge was desirable.

Six-foot-gauge railroads (6 ft [1,829 mm]) had developed a large regional following in New York State in the first part of the 19th century, due to the influence of the New York and Erie, one of the early pioneering railroads in America, chartered in 1832, with its first section opening in 1841. The builders and promoters decided that a six-foot track gauge would be needed for locomotives to be larger and more powerful than were in general use at the time, for pulling very large trains. Also the six-foot gauge provided greater stability, and the New York and Erie would operate passenger cars up to 11 feet (3.4 m) wide. Building westward from the Hudson River, it eventually reached Lake Erie, establishing a mainline longer than 400 miles (640 km) providing a shortcut to the American Midwest region from the New York City vicinity, and helping spawn a regional network of six-foot-gauge railroads almost exclusively within New York State.

Many early New York railways were Erie railroad-built branch lines, while others were independent railroads that wanted to partner and interchange with the Erie. These included the Walkill Valley, the Albany and Susquehanna (later part of the Delaware and Hudson); the Elmira, Jefferson & Canandaigua (later the Northern Central, becoming part of the Pennsylvania Railroad); the Delaware, Lackawanna and Western mainline (which also had a significant amount of trackage in Pennsylvania); predecessor lines of the New York and Oswego Midland (later the New York, Ontario, and Western); and the Canandaigua and Niagara Falls (later becoming part of the New York Central railroad's *Peanut Route* along the shoreline of Lake Ontario). However, by the late 1870s, the trend was inevitable, and conversion to standard gauge began, some lines first becoming "dual gauged" with the addition of a third running rail. Between 1876 and 1880, most of the remaining six-foot gauge trackage was converted.

In 1886, the railways in the Southern United States agreed to coordinate changing gauge on all their tracks. After considerable debate and planning, most of the southern rail network was converted from 5 ft (1,524 mm) gauge to 4 ft 9 in (1,448 mm) gauge, nearly the standard of the Pennsylvania Railroad, over two days beginning on 31 May 1886. Over a period of 36 hours, tens of thousands of workers pulled the spikes from the west rail of all the broad-gauge lines in the South, moved them 3 in (76 mm) east and spiked them back in place. The new gauge was close enough that standard-gauge equipment could run on it without difficulty. By June 1886, all major railroads in North America were using approximately the same gauge. The final conversion to true standard gauge took place gradually as track was maintained.

Some North American tram (streetcar) lines intentionally varied from standard gauge. This may have been to make the tram companies less tempting targets for takeovers by the steam railways (or competing trams companies), which would be unable to run their trains over the tram tracks.

A gauge in use is the Pennsylvania trolley gauge of 5 ft 2 1⁄2 in (1,588 mm), still used on the streetcars in New Orleans, and the Pittsburgh Light Rail system. This gauge was also used for the now defunct Pittsburgh Railways, West Penn Railways, and trams in Cincinnati. Similar 5 ft 2 1⁄4 in (1,581 mm) gauge is used in Philadelphia on the SEPTA routes, 15, 101 and 102, the Subway–Surface Trolleys and the SEPTA Market-Frankford. The Bay Area Rapid Transit (BART) system in the San Francisco Bay Area was opened in 1972 with 5 ft 6 in (1,676 mm) gauge, and thus requires all related equipment to be custom built. The system trackage has been extended multiple times since then, and new railcars have been ordered with this non-standard gauge.

The use of a non-standard gauge precludes interoperability of rolling stock on railway networks. On the British Great Western Railway the 7 ft 1⁄4 in (2,140 mm) gauge was supposed to allow high speed, but the company had difficulty with locomotive design in the early years, losing much of the advantage, and rapid advances in permanent way and suspension technology allowed standard-gauge speeds to approach broad-gauge speeds within a decade or two. On the 5 ft 3 in (1,600 mm) and 5 ft 6 in (1,676 mm) gauges, the extra width allowed bigger inside cylinders and greater power, a problem solvable by using outside cylinders and higher steam pressure on standard gauge. In the end, the most powerful engines on standard gauge in North America and Scandinavia far exceeded the power of any broad-gauge locomotive.

## Gauges

### 5 ft and 1520 mm Gauge

This gauge was first used in the United Kingdom and the United States before it became the standard gauge for most railways in the former Soviet Union.

Russian gauge or CIS gauge 1,520 mm (4 ft 11 $^{27}/_{32}$ in) is the second most widely used gauge in the world, and spans the whole of the former Soviet Union/CIS bloc including the Baltic states and Mongolia. Finland uses 1,524 mm (5 ft). The difference is clearly lower than the tolerance margin, so through running is feasible. Care must be taken when servicing international trains because the wear profile of the wheels differs from that of trains that run on domestic tracks only.

When the Finnish rail network was founded in 1862, Finland was the Grand Duchy of Finland, a region of Imperial Russia. The first border crossing railway to Russia was opened in 1870, while the first to Sweden was not until 1919, so railways were built to the broad Russian track gauge of 1,524 mm (5 ft). In Russia, this gauge was re-standardized to 1,520 mm (4 ft 11 $^{27}/_{32}$ in) during the 1960s. Finland retained the original gauge with no re-standardisation.

### 5 ft 3 in Gauge

Irish 5 ft 3 in broad gauge tracks.

As part of the railway gauge standardisation considered by the United Kingdom Parliamentary Gauge Commission, Ireland was allocated its own gauge, Irish gauge. Ireland then had three gauges, and the new standard would be a fourth.

The Irish gauge of 1,600 mm (5 ft 3 in) is used in Ireland and parts of Australia and Brazil. A problem with the Irish Gauge in Australia is that it is only about 170 mm (6.5 in) wider than the standard gauge used in other parts of Australia, principally New South Wales. Therefore, it is not considered advisable to use a third rail to allow dual-gauge operation on mainline sections of track, because of the danger of material lodging between the two rails. There has been argument for well over a century about the practicability of third rail operation, and numerous devices have been promoted to overcome the problem, especially at turnouts, including the “Brennan Switch”.

## Iberian Gauge

As finally established, the Iberian gauge of 1,668 mm (5 ft 5 $^{21}/_{32}$ in) is a compromise between the similar, but slightly different, gauges first adopted as respective national standards in Spain and Portugal in the mid-19th century. The main railway networks of Spain were initially constructed to a 1,672 mm (5 ft 5 $^{13}/_{16}$ in) gauge of six Castilian feet. Those of Portugal were initially built in standard gauge, but by 1864 were all converted to a 1,664 mm (5 ft 5 $^{1}/_{2}$ in) gauge of five Portuguese feet – close enough to allow interoperability in practice. The new high-speed network in Spain and Portugal uses standard gauge. The dual-gauge high-speed train RENFE Class 130 can change gauge at low speed without stopping.

## 5 ft 6 in Gauge

The 5 ft 6 in (1,676 mm) gauge was first used in Great Britain in Scotland for two short, isolated lines, the Dundee and Arbroath Railway and the Arbroath and Forfar Railway (1838-). Both the lines were subsequently converted to standard gauge and connected to the Scottish rail network.

Later this gauge was adopted as a standard for many British colonies such as Province of Canada and British India.

In 1851, the 5 ft 6 in (1,676 mm) broad gauge was officially adopted as the standard gauge for the Province of Canada, becoming known as the *Provincial gauge*, and government subsidies were unavailable for railways that chose other gauges. In the 1870s, mainly between 1872 and 1874, Canadian broad-gauge lines were changed to standard gauge to facilitate interchange and the exchange of rolling stock with American railways. Today, all Canadian railways are standard-gauge.

In US, this gauge was adopted for many lines, but soon fell out in favour of standard gauge. Today, only California's Bay Area Rapid Transit (BART) uses this gauge.

In British India, some standard gauge freight railways were built in initial period, though they were dismantled later. Later, in 1850s, the gauge of 5 ft 6 in (1,676 mm) was adopted as standard for the nationwide network. Attempts to economize on the cost of construction led to the adoption of 1,000 mm (3 ft 3 $^{3}/_{8}$ in) gauge and then 2 ft 6 in (762 mm) and 2 ft (610 mm) narrow gauges for many secondary and branch lines. In the later part of 20th century, due to interchangeability and maintenance issue, the railways in each of the countries in the Indian Subcontinent began to convert all metre-gauge and narrow-gauge lines to this gauge. Today, the nationwide rail network in Pakistan, Sri Lanka and Nepal is entirely on this gauge, whereas India, under Project Unigauge, and Bangladesh are still undergoing gauge conversion.

This gauge is the widest gauge in regular passenger use in the world.

## Broader Gauges

Some industrial applications require still broader gauges, such as:

- Large telescopes and telescope arrays.
- Rocket launchers – The European Space Agency, Russian Federal Space Agency, NASA

and SpaceX use double-track railway to move rockets and supporting equipment at launch sites (with exceptions such as the US Apollo program and Space Shuttles, which used caterpillar tracks on a river stone roadbed because other solutions could not support the loads required).

- Gantry crane for unloading cargo from ships, loading cargo onto ships and for constructing ships.
- Funiculars, Incline elevator and Canal inclined plane, for transportation on sloped way.

## Iberian-gauge Railways

Iberian-gauge railways is the name given to the railways using track gauge of 1,668 mm (5 ft 5 $^{21}/_{32}$ in), most extensively used by the railways of Spain and Portugal. This is the second-widest gauge in regular use anywhere in the world.

As finally established in 1955, the Iberian gauge is a compromise between the similar, but slightly different, gauges adopted as respective national standards in Spain and Portugal in the mid-19th century. The main railway networks of Spain were initially constructed to a 1,672 mm (5 ft 5 $^{13}/_{16}$ in) gauge of six Castilian feet. Those of Portugal were instead built to a 1,435 mm (4 ft 8 $^{1}/_{2}$ in) and later railways to a 1,664 mm (5 ft 5 $^{1}/_{2}$ in) gauge of five Portuguese feet – close enough to allow interoperability with Spanish railways.

## Standard Gauge

Since the beginning of the 1990s new high-speed passenger lines in Spain have been built to the international standard gauge of 1,435 mm (4 ft 8 $^{1}/_{2}$ in), to allow these lines to link to the European high-speed network. Although the 22 km from Tardienta to Huesca (part of a branch from the Madrid to Barcelona high-speed line) has been reconstructed as mixed Iberic and standard gauge, in general the interface between the two gauges in Spain is dealt with by means of gauge-changing installations, which can adjust the gauge of appropriately designed wheelsets on the move.

Plans exist to convert more of the Iberian-gauge network in Spain and Portugal to standard gauge, an indication of which is the use, on several stretches of recently relaid broad-gauge track, of concrete sleepers pre-drilled with additional bolt holes allowing for repositioning of one rail to adjust the track to standard gauge (or to dual gauge) or the narrowing of the gauge by moving both rails closer together maintaining the perfect alignment of the loading gauge.

## Causes for the Difference from Standard Gauge

A commonly cited reason for the adoption of this gauge, broader than that which became the international standard, is that it was intended to hinder any use of the railroad in any of the French invasion attempts. This is possible, but the 19 January 1845 governmental decree does not state any reason, simply the gauge itself. It is known that the government entrusted such decisions to a committee, whose main work was done by civil engineers Juan Subercase and Calixto de Santa Cruz: both good civil engineers but with little experience in railroads. Their judgement, made because of their profession and lack of experience in railroads, was that since the Spanish landscape was more mountainous than that of England and France, a wider gauge would make more powerful locomotives possible, with larger boilers and, more importantly, larger fireboxes. This was obviously a miscalculation; in other areas of Spain, narrower gauges were used in the most mountainous areas, since Iberian gauge civil engineering works would have been more expensive. One example of this is the 1,000 mm (3 ft 3 $^{3}/_{8\text{ in}}$) gauge FEVE.

## Similar Gauges and Compatibility

The Indian gauge (1,676 mm or 5 ft 6 in) is closely similar, with only 8 mm ($^{5}/_{16}$ in) difference, and allows compatibility with the rolling stock. For example, in recent years Chile and Argentina have bought second hand Spanish/Portuguese Iberian-gauge rolling stock. 1,668 mm trains can run on 1,676 mm gauge without adaptation, but for better stability in high-speed running a little wheelset adjustment may be required (for example, the Russian-Finnish train Allegro is built for a 1,522 mm track gauge, intermediate between the Russian 1,520 mm and the Finnish 1,524 mm). Backward compatibility—1,676 mm trains on 1,668 mm gauge—is possible, but no examples and data exist. Due to the narrower gauge, a strong wear of wheelsets may occur without adjustment.

## Narrow-gauge Railway

A narrow-gauge railway is a railway with a track gauge narrower than standard 1,435 mm (4 ft 8 $^{1}/_{2}$ in). Most narrow-gauge railways are between 600 mm (1 ft 11 $^{5}/_{8}$ in) and 1,067 mm (3 ft 6 in).

Since narrow-gauge railways are usually built with tighter curves, smaller structure gauges, and lighter rails, they can be less costly to build, equip, and operate than standard- or broad-gauge railways (particularly in mountainous or difficult terrain). Lower-cost narrow-gauge railways are often built to serve industries and communities where the traffic potential would not justify the cost of a standard- or broad-gauge line.

Narrow-gauge railways have specialized use in mines and other environments where a small structure gauge necessitates a small loading gauge. They also have more general applications. Non-industrial, narrow-gauge mountain railways are (or were) common in the Rocky Mountains of the United States and the Pacific Cordillera of Canada, Mexico, Switzerland, the former Yugoslavia, Greece, and Costa Rica. In some countries, narrow gauge is the standard; Japan, Indonesia, Taiwan, New Zealand, South Africa, and the Australian states of Queensland, Western Australia and Tasmania have a 3 ft 6 in (1,067 mm) gauge, and Malaysia and Thailand have metre-gauge railways. Narrow-gauge trams, particularly metre-gauge, are common in Europe.

A narrow-gauge railway is one where the distance between the inside edges of the rails is less than 1,435 mm (4 ft 8 $^{1}/_{2}$ in). Historically, the term was sometimes used to refer to standard-gauge railways, to distinguish them from broad-gauge railways, but this use no longer applies.

## Advantages

Narrow-gauge railways usually cost less to build because they are usually lighter in construction, using smaller cars and locomotives (a smaller loading gauge), smaller bridges and tunnels (a smaller structure gauge), and tighter curves. Narrow gauge is often used in mountainous terrain, where engineering savings can be substantial. It is also used in sparsely populated areas where the potential demand is too low for broad-gauge railways to be economically viable. This is the case in parts of Australia and most of Southern Africa, where poor soils have led to population densities too low for standard gauge to be viable.

For temporary railways which will be removed after short-term use, such as logging, mining or large-scale construction projects (especially in confined spaces, such as the Channel Tunnel), a narrow-gauge railway is substantially cheaper and easier to install and remove. Such railways have almost vanished, however, due to the capabilities of modern trucks.

In many countries, narrow-gauge railways were built as branch lines to feed traffic to standard-gauge lines due to lower construction costs. The choice was often not between a narrow- and standard-gauge railway, but between a narrow-gauge railway and none at all.

## Disadvantages and Solutions

### Interchangeability

Narrow-gauge railways cannot freely interchange rolling stock (such as freight and passenger cars) with the standard- or broad- gauge railways with which they link, and the transfer of passengers and freight require time-consuming manual labour or substantial capital expenditure. Some bulk commodities, such as coal, ore, and gravel, can be mechanically transshipped, but this is time-consuming, and the equipment required for the transfer is often complex to maintain.

If rail lines with other gauges coexist in a network, in times of peak demand it is difficult to move rolling stock to where it is needed when a break of gauge exists. Sufficient rolling stock must be available to meet a narrow-gauge railway's peak demand (which might be greater in comparison to a single-gauge network), and the surplus equipment generates no cash flow during periods of low demand. In regions where narrow gauge forms a small part of the rail network (like Russia's Sakhalin Railway), extra money is needed to design, produce or import narrow-gauge equipment.

Solutions to interchangeability problems include bogie exchanges, a rollbock system, variable gauge, dual gauge or gauge conversion.

### Growth Potential

Another problem commonly faced by narrow-gauge railways is that they lack the space to grow; their cheap construction was engineered only for initial traffic demands. Many narrow-gauge railways were impractical to improve; speeds and loads hauled could not increase, so traffic density was limited. In Japan, a few narrow-gauge lines have been upgraded to standard-gauge mini-shinkansen to allow through service by standard-gauge high-speed trains; due to the alignment and minimum curve radius of those lines, however, the maximum speed of the through service is the same as the original narrow-gauge line. If a narrow-gauge line is built to a higher standard, like Japan's proposed Super Tokkyū, this problem can be minimized.

If narrow-gauge rails are designed with potential growth in mind (or at the same standard as standard-gauge rails), obstacles to future growth would be similar to other rail gauges. For lines constructed to a lower standard, speed can be increased by realigning rail lines to increase the minimum curve radius, reducing the number of intersections or introducing tilting trains.

## Successful Railways

An Electric Tilt Train in Queensland. Unlike other states in Australia which use different gauges, Queensland's network is made up of 3 ft 6 in (1,067 mm)-gauge track.

The heavy-duty 3 ft 6 in (1,067 mm) narrow-gauge railways in Queensland, South Africa, and New Zealand demonstrate that if track is built to a heavy-duty standard, performance almost as good as a standard-gauge line is possible. Two-hundred-car trains operate on the Sishen–Saldanha railway line in South Africa, and high-speed Tilt Trains run in Queensland. Another example of a heavy-duty narrow-gauge line is Brazil's EFVM. 1,000 mm (3 ft 3 3⁄8 in) gauge, it has over-100-pound rail (100 lb/yd or 49.6 kg/m) and a loading gauge almost as large as US non-excess-height lines. The line has a number of 4,000-horsepower (3,000 kW) locomotives and 200-plus-car trains. In South Africa and New Zealand, the loading gauge is similar to the restricted British loading gauge; in New Zealand, some British Rail Mark 2 carriages have been rebuilt with new bogies for use by Tranz Scenic (Wellington-Palmerston North service), Tranz Metro (Wellington-Masterton service), and Transdev Auckland (Auckland suburban services).

### Fastest Trains

Narrow gauge's reduced stability means that its trains cannot run at speeds as high as on broader gauges. For example, if a curve with standard-gauge rail can allow speed up to 145 km/h (90 mph), the same curve with narrow-gauge rail can only allow speed up to 130 km/h (81 mph).

In Japan and Queensland, recent permanent-way improvements have allowed trains on 3 ft 6 in (1,067 mm) gauge tracks to exceed 160 km/h (99 mph). Queensland Rail's Electric Tilt Train, the fastest train in Australia and the fastest 3 ft 6 in (1,067 mm) gauge train in the world, set a record of 210 km/h (130 mph). The speed record for 3 ft 6 in (1,067 mm) narrow-gauge rail is 245 km/h (152 mph), set in South Africa in 1978.

A special 2 ft (610 mm) gauge railcar was built for the Otavi Mining and Railway Company with a design speed of 137 km/h. Curve radius is also important for high speeds: narrow-gauge railways allow sharper curves, which limits a vehicle's safe speed.

### Gauges

Many narrow gauges, from 15 in (381 mm) gauge and 1,435 mm (4 ft 8 1⁄2 in) gauge, are in present or former use. They fall into several broad categories:

### 4 ft 6 in Gauge

4 ft 6 in (1,372 mm) track gauge (also known as Scotch gauge) was adopted by early 19th-century railways, primarily in the Lanarkshire area of Scotland. 4 ft 6 1⁄2 in (1,384 mm) lines were also constructed, and both were eventually converted to standard gauge.

## 4 ft and 1200 mm gauge

### 4 ft (1,219 mm)

- City of Oxford Tramways Company
- Glasgow Subway
- Padarn Railway
- Bradford Corporation Tramways
- Keighley Tramways
- Wellington tramway system
- Saundersfoot Railway
- Derby Tramways Company
- Reading Corporation Tramways
- Barrow-in-Furness Tramways Company
- Darwen Corporation Tramways

### 1,200 mm (3 ft 11 $^1/_4$ in)

- Central Funicular
- Gardena Ronda Express
- Zagreb Funicular
- Rheineck–Walzenhausen mountain railway
- Schlossbergbahn (Freiburg)
- Appenzell Railways
- Stoosbahn

## 3 ft 6 in Gauge

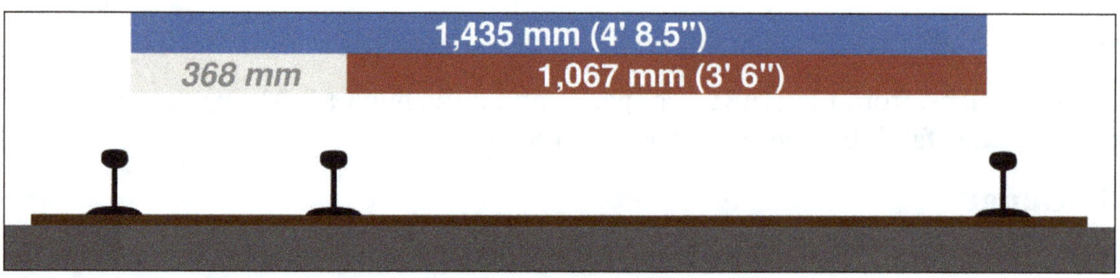


Comparison of 4 ft 8 $^1/_2$ in (1,435 mm) standard gauge (blue) and 3 ft 6 in (1,067 mm) (red) width; the difference is 14.5 in (370 mm), or about 26 percent of standard gauge.

1,067 mm (3 ft 6 in) between the inside of the rail heads, its name and classification vary worldwide and it has about 112,000 kilometres (70,000 mi) of track.

### Similar Gauges

- 1,055 mm (3 ft 5 $^1/_2$ in) in Algeria.
- 1,050 mm (3 ft 5 $^{11}/_{32}$ in) on the Hejaz railway in Israel, Jordan, Lebanon, Saudi Arabia and Syria; only a few lines survive.

### Metre Gauge and Italian Metre Gauge

As its name implies, metre gauge is a track gauge of 1,000 mm (3 ft 3 $^3/_8$ in). It has about 95,000 km (59,000 mi) of track.

According to Italian law, track gauges in Italy were defined from the centre of each rail rather than the inside edges of the rails. This gauge, measured 950 mm (3 ft 1 $^3/_8$ in) between the edges of the rails, is known as Italian metre gauge.

### 3 ft, 900 mm and Swedish Three Foot Gauge

The 3 ft (914 mm) gauge Disneyland Railroad.

There were a number of large 3 ft (914 mm) railroad systems in North America; notable examples include the Denver & Rio Grande and Rio Grande Southern in Colorado and the South Pacific Coast and West Side Lumber Co of California. 3 ft was also a common track gauge in South America, Ireland and on the Isle of Man. 900 mm was a common gauge in Europe. Swedish three-foot-gauge railways (891 mm or 2 ft 11 $^3/_{32}$ in) are unique to that country.

### 2 ft 9 in Gauge

A few railways and tramways were built to 2 ft 9 in (838 mm) gauge, including Nankai Main Line (later converted to 3 ft 6 in or 1,067 mm), Ocean Pier Railway at Atlantic City, Seaton Tramway (converted from 2 ft) and Waiorongomai Tramway.

### 800 mm, 2 ft 6 in, Bosnian and 750 mm Gauge

800 mm (2 ft 7 $^1/_2$ in) gauge railways are commonly used for rack railways. Imperial 2 ft 6 in (762 mm) gauge railways were generally constructed in the former British colonies. 760 mm Bosnian gauge and 750 mm railways are predominantly found in Russia and Eastern Europe.

## Gauges between 2 ft and 2 ft 6 in

Gauges such as 2ft 3in (686mm), 2ft 4in (711mm) and 2ft 4.5in (724mm) were used in parts of the UK, particularly for railways in Wales and the borders, with some industrial use in the coal industry. Some sugar cane lines in Cuba were 2ft 3.5in (698.5mm).

## 2 ft and 600 mm Gauges

The 1 ft 11 ½ in (597 mm) gauge Ffestiniog Railway.

2 ft (610 mm) gauge railways were generally constructed in the former British colonies. 1 ft 11 ¾ in (603 mm), 600 mm (1 ft 11 ⅝ in) and 1 ft 11 ½ in (597 mm) were used in Europe.

## Minimum Gauge

Gauges below 1 ft 11 ½ in (597 mm) were rare. Arthur Percival Heywood developed 15 in (381 mm) gauge estate railways in Britain and Decauville produced a range of industrial railways running on 500 mm (19 ¾ in) and 400 mm (15 ¾ in) tracks, most commonly in restricted environments such as underground mine railways, parks and farms, in France. Several 18 in (457 mm) gauge railways were built in Britain to serve ammunition depots and other military facilities, particularly during World War I.

## Minimum-gauge Railway

Minimum-gauge railways have a gauge of most commonly 15 in (381 mm), 400 mm (15 ¾ in), 16 in (406 mm), 18 in (457 mm), 500 mm (19 ¾ in) or 20 in (508 mm). The notion of minimum-gauge railways was originally developed by estate railways and the French company of Decauville for industrial railways, mining, and farming applications.

The term was originally conceived by Sir Arthur Percival Heywood who used it in 1874 to describe the principle behind his Duffield Bank Railway, specifically its 15 in (381 mm) gauge, distinguishing it from a "Narrow Gauge" railway. Having previously built a small railway of 9 in (229 mm) gauge, he settled on 15 in (381 mm) as the minimum that he felt was practical.

In general, minimum-gauge railways maximize their loading gauge, where the dimension of the equipment is made as large as possible with respect to the track gauge while still providing enough stability to keep it from tipping over. Minimum-gauge railways allowed for ease of mobility on battlefields, mines, and other restricted environments.

A number of 18 in (457 mm) gauge railways were built in Britain to serve ammunition depots and other military facilities, particularly during the First World War.

In France, Decauville produced a range of portable track railways running on 400 mm (15 $^{3}/_{4\text{ in}}$) and 500 mm (19 $^{3}/_{4\text{ in}}$) tracks, most commonly in restricted environments such as underground mine railways, parks and farms.

During World War II, it was proposed to expedite the Yunnan–Burma Railway using 400 mm (15 $^{3}/_{4\text{ in}}$) gauge, since such a small gauge can have the tightest of curves in difficult terrain.

### Distinction between Ridable Miniature Railway and Minimum-gauge Railway

The major distinction between a miniature railway (USA: 'riding railroad' or 'grand scale railroad') and a minimum-gauge railway is that miniature lines use models of full-sized prototypes. There are miniature railways that run on gauges as wide as 2 ft (610 mm), for example the Wicksteed Park Railway. There are also rideable miniature railways running on extremely narrow tracks as small as 10 $^{1}/_{4\text{ in}}$ (260 mm) gauge, for example the Rudyard Lake Steam Railway. Around the world there are also several rideable miniature railways open to the public using even narrower gauges, such as 7 $^{1}/_{4\text{ in}}$ (184 mm) and 7 $^{1}/_{2\text{ in}}$ (190.5 mm).

Generally minimum-gauge railways have a working function as estate railways, or industrial railways, or providers of public transport links; although most have a distinct function in relation to tourism, and depend upon tourism for the revenue to support their working function.

## Dual Gauge

A dual gauge railway is a track that allows the passage of trains of two different track gauges. It is sometimes called a "mixed gauge" track. A dual gauge track consists of three rails. There will be two vital rails ("gauge" rails), one for each gauge close together and a third rail, a "common" rail further away. Sometimes, four rails are required using two outer and two inner rails to create the dual gauge. Dual gauge is not to be confused with a "third rail" (the electric current traction rail) or "check or guard rails".

Rail gauge, the distance between the inner surfaces of the heads of travel rails, is an important specification of a railway. Rail tracks and wheel bogies must be built to the same gauge within an engineering tolerance of 13 mm (0.5 in). If the correct gauge is not achieved, the train will fall off the track and not be able to pass switches and crossovers.

Dual gauge trains can use low level platforms because their carriages rest higher than the platform. In the case of three rails and high platforms, one gauge may be too close or too far away, depending on the position of the common rail. Another option at platforms is to construct separate tracks, one for each gauge.

### Tolerance

If the difference between two rail gauges is small enough, i.e. within each other's tolerances, then it is possible for them to operate the same rolling stock.

At the Finland–Russian border the Finnish railway gauge is 1524 mm and the Russian gauge is 1520 mm. When the Soviet Union changed the gauge of its railways in Russia in the 1970s to 1520 mm, this did not result in a break of gauge and no track conversion work was done. The change in gauge was a redefinition of the way tolerances were measured. Both railways remained well within each other's tolerance and can run the same rolling stock.

However, being within a tolerance in gauge does not always mean that two different system can successfully operate the same rolling stock. For example, the MTR (Mass Transit Railway) in Hong Kong 1,432 mm (4 ft 8 $^{3}/_{8}$ in) Electric multiple units (EMU) may run on Kowloon-Canton Railway (KCR) 1,435 mm (4 ft 8 $^{1}/_{2}$ in) rails but would need a locomotive or a KCR EMU pulling due to the difference in electrification voltages. Dual voltage and frequency EMUs are the other solution.

## Break of Gauge

"Break of gauge" occurs when rails of different gauges outside the normal tolerance meet. Passengers and freight must transfer between trains, or rolling stock must be lifted and the bogies refitted for the new gauge. Avoiding break of gauge reduces costs and allows infrastructure such as platforms, bridges and tunnels to be shared.

## Gauge Conversion

Railway operators may change from one gauge to another via a period of dual gauge operations. For example, the Great Western Railway (GWR) made a conversion from a 7-foot broad gauge to the standard gauge via a period of dual gauge operations across its network. New GWR rolling stock and locomotives of that time were built to accommodate the change.

Where rails are too light for the loads of broader-gauge railcars, dual gauge rails may not be feasible. In this case, heavier rails are installed.

## Configurations

One common running rail and two other outer rails provide a dual gauge. In dual gauge lines, railroad switches (points) are more complex. Trains must be safely signalled on both of the gauges. Track circuits and mechanical interlocking must also operate on both gauges. Another feature is that the wear and tear of the common rail is greater than the two other outer rails.

## Three Rails

Three rail dual gauge.

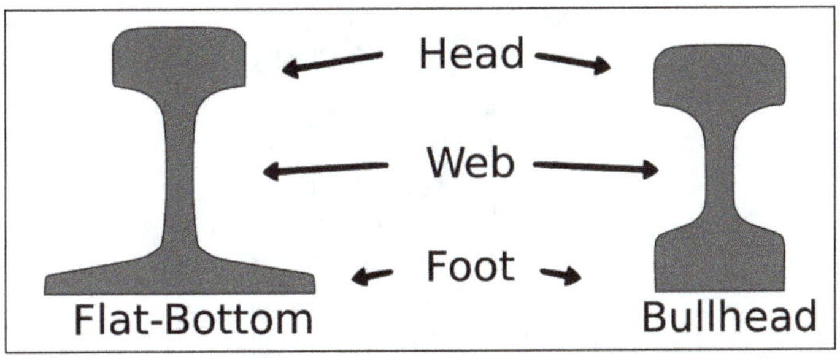


Cross-sections of flat-bottomed rail which can rest directly on the sleepers, and bullhead rails which sit in chairs.

Dual gauge track with three rails must provide a difference between the gauges at least as wide as the foot of the rail. This is to ensure there is room for rail fastening hardware such as spikes and clips.

Functional pairing of gauges include: 1,435 mm (4 ft 8 ½ in) standard gauge and 1,676 mm (5 ft 6 in); 3 ft (914 mm) and 3 ft 6 in (1,067 mm); and 5 ft (1,524 mm) and 6 ft (1,829 mm). 1,435 mm (4 ft 8 ½ in) standard gauge and 1,600 mm (5 ft 3 in) can also be dual gauged, albeit with lighter, narrow footed rails. An example of this type of pairing is seen in Victoria, Australia.

## Four Rails

Gauges which are too close to function in a three rail arrangement include 1,000 mm (3 ft 3 ⅜ in) metre gauge and 3 ft 6 in (1,067 mm) (common in Africa); 1,000 mm (3 ft 3 ⅜ in) and 3 ft (914 mm) (common in South America); and 1,435 mm (4 ft 8 ½ in) and 1,524 mm (5 ft), and 1,524 mm (5 ft) and 1,676 mm (5 ft 6 in). The last combination is common in Afghanistan, Central Asia, northern, central and eastern Europe, Russia, North America, Iran, and China. In Europe, it was of strategic importance in World War II.

In these cases, a gauntlet track which uses four rails is constructed. An example of this is seen at the Rail Baltica project which aims to connect central and northern Europe by rail.

Four rails may also be used where a co-location of track centres of the two gauges is needed. This might occur in tight tunnels or past platforms. An example is seen at the Roma Street railway station in Brisbane, Queensland, Australia. There, both three rail and four rail dual gauge systems are used between 1,435 mm (4 ft 8 ½ in) and 1,067 mm (3 ft 6 in) gauges.

1524 mm and standard gauge on a bridge between Haparanda (Sweden) and Tornio (Finland).

Four rail track with a complicated switch on the Chemin de Fer de la Baie de Somme.

## Triple Gauge

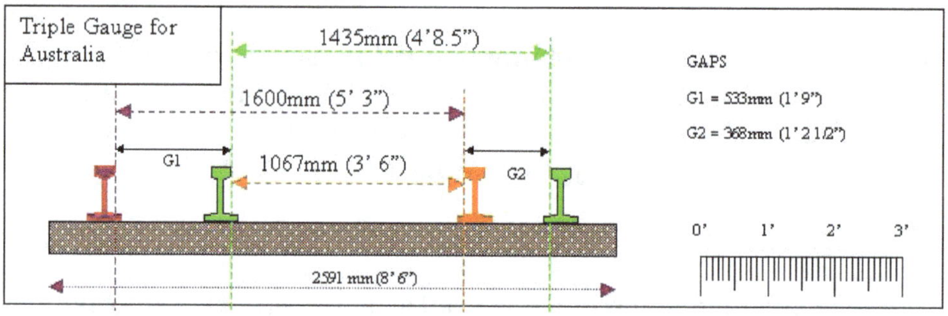


Triple gauge used in the station yards at Gladstone and Peterborough.

Break of gauge occurs at some triple gauge stations. The triple gauge was used in rail yards where trains operate at low speeds. Thus, if required, light rail could be used to space the rails closely together. Light rail was not used at the Niagara Falls Suspension Bridge and it would not be used for main line operation at high speeds.

## Quadruple Gauge

Quadruple gauge rail tracks at the Alan Keef locomotive works at Lea.

Within a works facility or maintenance yard, tracks consisting of four or more separate gauges may be used. At Alan Keef in Lea, Herefordshire a short section of line uses four rails to allow locomotives of 2 ft (610 mm), 2 ft 6 in (762 mm), 3 ft (914 mm) and 3 ft 6 in (1,067 mm) gauges to enter the works.

The National Railway Museum, Port Adelaide in Adelaide, Australia has the three main-line gauges and a 18 in (457 mm) gauge Heritage railway line.

In the 1960s and 1970s, the Electro-motive diesel (EMD) plant in McCook, Illinois maintained a small amount of multi-gauge track with up to seven parallel rails in order to support the wide variety of export locomotives they produced. This track did include a turnout splitting a standard gauge track from the seven-rail track. It required eight frogs in a row.

## Switches

Dual gauge switches (also known as "turnouts" or "points"), where both gauges have a choice of routes, are more complicated than those where two gauges separate or single gauge switches. The train must move through a dual gauge switch very slowly. The ends of dual gauge switches are easier to design if their operation is electrical rather than mechanical.

If two gauges are similar in width, the switch they use will have many small pieces that are difficult to support. The switch will also be limited in speed. The difference between the gauges should be 50 mm greater than the width of the base of the rails. In a rail yard, weak dual gauge switches are avoided by separating the gauges and using, single gauge switches, and dual gauge diamond crossings. Gauge splitters are used for trains of a single gauge. The fixed type have no moving parts and trains move through them slowly. Power operated gauge splitters are operated like ordinary switches.

## Other Configurations

### Separate Gauge

Dual gauge lines are separated by building two tracks, one of each gauge, side by side. Whether a dual gauge line remains depends on the volume of rail traffic it carries and also its location, for example across a bridge or through a tunnel. Separated lines share infrastructure such as signal boxes and signallers.

In Victoria (Australia), on the Melbourne to Geelong line 80 km (50 mi), a single 1,435 mm (4 ft 8 1⁄2 in) standard gauge line runs parallel to the double track broad gauge.

Prior to 1941, the 58 km (36 mi) Yogyakarta – Surakarta line in Java, in the Dutch East Indies, had a single 3 ft 6 in (1,067 mm) line in parallel with a dual gauge 1,435 mm (4 ft 8 1⁄2 in) and a 3 ft 6 in (1,067 mm) line.

In 1960, in Western Australia, the Perth to Northam line was to be a 3 ft 6 in (1,067 mm) narrow gauge track running in parallel with a 1,435 mm (4 ft 8 1⁄2 in) line. Planners then realised a double dual gauge line would increase capacity.

The Uzhhorod–Košice broad-gauge track in Slovakia runs in parallel to a double track 1,435 mm (4 ft 8 1⁄2 in) line.

### Overlapping Gauges

Railway of one gauge may be extended into territory that uses another gauge. This may avoid transhipment. For example, a 1,524 mm (5 ft) gauge line runs from an iron ore mine in Ukraine to a steelworks in Slovakia. A portion of four rail dual gauge line (1435 mm and 1520 mm) runs from Slovakia to Romania (both standard gauge) through Ukraine (Russian gauge).

Flat crossing of Lisbon trams' 900 mm (2 ft 11 $^{7}/_{16}$ in) gauge track with regular freight track in Iberian gauge.

### Other Procedures

Transporter wagons, transporter trucks, and rollbocks are used to carry vehicles built for one gauge on a line with a different gauge. They can manage a trainload at a time. Bridges and tunnels must be one metre higher than usual.

Bogie exchange systems lift the railroad car while trucks or bogies are changed for a different gauge. This system is not suitable for four-wheeled wagons. The Ramsey car transfer apparatus is another way to change bogies.

Adjustable gauge equipment or variable gauge axles, allow a wheel gauge to be altered. Transshipment procedures move containers and people from one train to another.

### Model Railways/Miniature Ridable Railways

Tillig make dual gauge model railway track.

Many ridable miniature railways utilise dual gauge, often with models of the same prototype built to different scales to match the gauge.

## Advantages and Disadvantages of Different Track Gauges

- Pros: Tighter turning radius, lower cost, less demanding right-of-way and construction.
- Cons: Lower speed, less stability, less load carrying capacity.
- Pros: Higher speed, stability and capacity.
- Cons: Wider turning radius, higher cost, more demanding right-of-way and construction.

One generally wants higher speed, greater stability and larger capacity, and one wants lower cost and tighter turning radius, but there is an inverse relationship between these factors.

In addition, there are other constraints, such as the load-carrying capacity of axles, which can be problematic with an excessively wide gauge.

Narrow gauge railways usually cost less to build because they are usually lighter in construction, using smaller cars and locomotives (smaller loading gauge), as well as smaller bridges, smaller tunnels (smaller structure gauge) and tighter curves. Narrow gauge is thus often used in mountainous terrain, where the savings in civil engineering work can be substantial.

It is also used in sparsely populated areas, with low potential demand, and for temporary railways that will be removed after short-term use, such as for construction, the logging industry, the mining industry, or large-scale construction projects, especially in confined spaces.

Broader gauge railways are generally more expensive to build and require wider curves, but are able to handle heavier and faster traffic.

There is no single perfect gauge, because different environments and economic considerations come into play.

A narrow gauge is superior if one's main considerations are economy and tight curvature (e.g. Japan relies on the 3'6″ gauge due to its mountainous terrain).

For direct, unimpeded routes with high traffic, a broad gauge may be preferable (which is why the BART train system uses Indian/5'6″ gauge). The Standard, Russian, and 4'6″ gauges are designed to strike a reasonable balance between these factors.

In addition to the general trade-off, another important factor is standardization.

Once a standard has been chosen, and equipment, infrastructure, and training calibrated to that standard, conversion becomes difficult and expensive.

This also makes it easier to adopt an existing standard than to invent a new one.

This is true of many technologies, including railroad gauges. The reduced cost, greater efficiency, and greater economic opportunity offered by the use of a common standard explains why a small number of gauges predominate worldwide.

## References


- Rail-gauges: railsystem.net, Retrieved 7 July, 2019
- "New Swift international train makes first journey with passengers from Moscow to Berlin". Rzd.ru. Russian Railways. Retrieved 31 March 2017
- Break-of-guage-rail-gauges: worldwiderails.com, Retrieved 8 August, 2019
- "Operational Requirements" (PDF). Development of the Trans-Asian Railway: Trans-Asian Railway in the North-south Corridor, Northern Europe to the Persian Gulf. 2001. Pp. 49–58. ISBN 978-92-1-120099-7
- Advantages-disadvantages-different-track-gauges: transportday.com.ng, Retrieved 9 January, 2019
- Bruzek, Radim; Trosino, Michael; Kreisel, Leopold; Al-Nazer, Leith (2015). "Rail Temperature Approximation and Heat Slow Order Best Practices". 2015 Joint Rail Conference. pp. V001T04A002. doi:10.1115/JRC2015-5720. ISBN 978-0-7918-5645-1

# 4

# Railway Signaling

Railway signaling is used to direct railway traffic and to keep the trains clear of each other. Some of the infrastructural items where signals are installed are points, moveable bridges and level junctions. This chapter has been carefully written to provide an easy understanding of the importance of railway signaling as well as the different types of signals.

## Railway Signal and its Types

A railway signal is a type of signal that conveys instructions or provides advance warning of instructions regarding the driver's authority to proceed. There are two types of signals, permissive and absolute. You can tell if a signal is permissive or absolute by looking at it. The rules vary by railroad, but permissive signals usually have a number plate on the base of the mast on which they're mounted, or the letter "I". Absolute signals do not have the number plate or will have the letter "A". The location of the signal is also often a clue to its type.

Signals have two purposes, protection and control. Permissive signals only provide protection. Absolute signals provide both control and protection. Functionally, the basic difference between permissive and absolute signals is the most restrictive type of indication each can display.

A red permissive signal means stop and proceed. After stopping (which is no longer required on some railroads) the train can proceed at restricted speed until a more favorable signal is reached. (Restricted speed is usually 15 or 20 mph, being prepared to stop upon sighting any obstruction or problem.)

But a red absolute signal means stop - and stay stopped. The only way to pass a red absolute signal is with the dispatcher's verbal permission or, in rare circumstances, under the protection of a flagman.

### Permissive Signals and Automatic Block Signaling

Permissive signals respond to track occupancy. The invention of the track circuit in 1872 by William Robinson permitted automatic block signals. Most simply, a low voltage battery current runs up one rail and down the other. As long as current flows from the battery through one rail to the relay and back through the other rail, the relay remains energized and routes current from another battery to the green lamp of a clear signal. But when a train is present, the circuit is short circuited by the steel wheels and axles of the cars; a broken rail or an open switch will also break the circuit. As a result the relay is opened and the signal displays red. The system is arranged so the failure of any component results in a red indication or a dark signal (which must be interpreted as the most restrictive aspect that signal can display).

The Automatic Block Signal (ABS) system permitted much shorter blocks than the manual block system and had the effect of increasing line capacity. The distant signal proved just as useful for the automatic block system as for manual block. It became common practice with lower-quadrant semaphores to place the home signal for the immediate block and the distant signal for the next block on the same mast.

An automatic block signal system is composed of a series of block signals. Except where a current of traffic has been established on a track and Rule 251 movement authority is in effect, permissive signals and ABS do not authorize the use of, or movement on, the main track. Their role is that of protective overlay: an added layer of protection placed on the operation. If a train violates its movement authority, it and a conflicting movement will both soon encounter red permissive signals, resulting in much embarrassment, but not an accident.

As a practical matter, Centralized Traffic Control (CTC) territory always includes ABS as a protective overlay. Territory controlled by Track Warrrant Control (TWC), Direct Traffic Control (DTC), or Form D Control System (DCS) may or may not include an ABS overlay; this commonly depends on the traffic volume of the particular line. Herein lies a unique circumstance: A train can be at a green signal in TWC territory and be unable to proceed, unless it also has a track warrant to do so. Likewise it can be facing a red permissive signal and yet proceed at restricted speed if it holds the proper TWC (or DTC, DCS, etc.) movement authority.

Block signals are often "approach lit" to conserve bulbs and electricity. They remain dark until the approach of an oncoming train is detected, and are illuminated only when a train is in the block that the signal faces. The circuitry is more complicated, but lamps last longer and power consumption is less.

## Absolute Signals and Interlocking Signals

Absolute signals are most often used at interlockings or in CTC territory where it is desired to control when, and if, a train passes a signal. Absolute signals are found at railroad crossings and drawbridges, and at places where trains pass over powered switches, such as at the ends of sidings or at main track crossovers. For two lightly trafficked rail lines in a remote area, stop signs may be sufficient as a type of absolute signal. However, just as four-way stop signs are an inefficient way to control the intersection of two roads streets, stop signs are not efficient on rail routes with heavy traffic.

When a train enters a siding and stops at the red absolute signal at the far end, the dispatcher obviously does not want it to "stop and proceed" down the line toward an oncoming train.

Usually the signals are manually controlled from a remote office. Outside CTC limits, a sophisticated ABS system that can detect train direction also will use them to protect movement between sidings. Such a system is termed an Absolute Permissive Block System.

Absolute signals usually "wear two hats" by also functioning as block signals to protect trains from colliding with each other, although they do not permit trains to pass a red signal after stopping. For example, in single-track CTC territory absolute signals protect the switches at the ends of sidings; permissive signals are placed at 2-3 mile intervals along the single track.

Absolute signals guard each end of an interlocking plant.

## Reading Signals: Aspects and Signal Indications

The lights displayed by a signal have three properties: aspect, name, and indication. The aspect describes what is displayed, e.g., red over green. The name is the formal name classification for that signal, e.g., diverging clear. The indication is the instruction conveyed by the signal.

A train crew seeing a signal with a yellow aspect will understand they are nearing an approach signal, which conveys an indication that might mean "proceed prepared to stop before your train passes the next signal; trains exceeding 35 mph must immediately reduce to that speed." (There's more to it than simply "the next signal is red").

All railroads use the common red, yellow, and green aspects, although the indication of the yellow (approach) signal varies from road to road. Many other color combinations are used. Some are common, others are unique to a specific railroad.

The most constant signals and the easiest to understand are single-head block signals. Red means stop; green means proceed, and yellow means caution or approach, usually indicating that the next signal is red. Four-indication block signaling usually uses either a flashing yellow aspect or two signal heads showing yellow-over-green or yellow-over-yellow to indicate two blocks clear ("advance approach" or "approach medium").

Block signals are normally clear. About all they can tell a train-watcher is that a train has just gone past or, on a single-track line, that there is something coming (from behind the observer), though not necessarily in the block-most automatic block systems are arranged so that a moving train will set signals at stop in front of it as far as the next siding.

The two- and three-head signals found at interlockings are more complex. All red means stop. That's universal. For other aspects, the general principle is that green on top means main route or normal speed, and red over green or yellow indicates diverging route or medium speed.

Interlocking signals are normally set at stop. Any less restrictive indication means that a route has been set up through the interlocking - in other words, something is coming. The preceding block signal usually serves as the distant signal for an interlocking. It will normally indicate caution; if it is clear, it indicates that a route has been set up through the interlocking.

The chance for each road to show its individuality comes with such things as flashing lights, white lights, blue lights, and dark signal heads.

Signal aspects, names, and indications once were shown in the rulebook, but because of rulebook consolidation and the increasing differences between roads, many railroads now show them in their employee timetable.

### Hardware: Different Types of Equipment

Over railroading's long history there have been many types of signals, using a variety of means to transmit their messages. Several designs are in use today.

### Ball Signals

The first U.S. signal installation was on the New Castle & Frenchtown Railroad in 1832. The NC&F signals were peach baskets suspended from masts by pulleys. By their position (high, middle, or low) and color (black or white), they conveyed information to trains and stations, whose staff observed them with telescopes.

With refinements, this "ball" type signal became widely used, its aspect for "proceed" giving us the stirring term "Highball!" The last ball signal in service is found at Whitefield, N.H.

The mid-19th century saw development of other manual signal types. Later, electricity was used to actuate signals powered by clockwork mechanisms. There were also electro-pneumatic types. All-electric designs came to be standard for automatic signals.

A Boston and Maine freight rolls past a ball signal in the "proceed" position at Coos Junction, N.H.

### Semaphore Signals

Perhaps the most classic of all railroad signals, the semaphore, came into general use in the 1860's. This type, based on an ancient form of communication that used a moveable "arm" or "blade," soon became by far the most common signal.

Various signal indications were conveyed by setting the blade in different positions. The problem of night visibility was addressed by using lamps in conjunction with the blade. Semaphores were constructed with castings in which were set red, yellow, and green lenses, which were moved in front of a single white light (oil at first, later electric) as the blade moved.

At first, the "lower-quadrant" semaphore design was used - the blade moved in the lower-right portion of an imaginary circle whose center was the blade's pivot point. Most lower-quadrant semaphores had two aspects: A horizontal (raised) blade meant "stop;" a blade hanging down about 60 degrees from horizontal meant proceed. Lower-quadrant blades had to be counterweighted so they would go to the horizontal position in case of failure of the controlling mechanism. Although there were three-position lower-quadrant semaphores, the two-position Union Switch & Signal "Style B" of 1898 (usually employing two arms) has come to be the best known.

Shortly after the turn of the century, the three-position (vertical, diagonal, horizontal) upper-quadrant semaphore was introduced. It had three aspects: horizontal for stop, 45 degrees above horizontal for caution, vertical for proceed. In 1908 this type became the standard for new installations. Such signals still exist on several roads. With the advent of light signals, semaphores came to be seen as inherently inferior, since they used two different aspects - blade position by day and light color by night - to convey their message. Also, their relatively numerous and heavy moving parts made them less attractive from a maintenance standpoint.

Close-up of an upper-quadrant semaphore signal.

## Color-light Signals

The need to provide signals in subway tunnels led to the development of color-light signals in the first decade of this century. While lights had been used for years with semaphores, their exclusive use in high-speed, main-line applications became feasible in 1914 with the perfection of concentrated filament lamps, giving satisfactory sighting distance in daylight with moderate power consumption.

Color-light signals consist of three bulbs shining through three lenses, or "roundels," one each red, yellow, and green. Typically, the lights are grouped vertically with green at the top, exactly opposite of the highway traffic signals that protect road intersections. Next most common is the triangular arrangement. Some roads, like Chicago & North Western, used a horizontal grouping.

Unlike semaphore lights, color-light signals are equipped with visors so sunlight reflections will not give false readings. Union Pacific fits its "tri-lights" with large hoods for this purpose.

Color-light signals grouped vertically protect Southern Railway's main line near Burke, Va.

## Searchlight Signals

The searchlight signal - which uses a single lamp with a system of movable lenses to project three colors - first appeared in 1920. Lens improvements in 1930 made signals with a 1-mile range and low power consumption possible, and the searchlight found widespread use.

A relay-type mechanism moves colored lenses in front of a white light, not unlike a semaphore. To achieve long range, the light beam is very intense and narrow, and must be precisely aimed for best visibility from approaching locomotive cabs.

To many, the searchlight's simple, clean appearance made it a perfect aesthetic complement to the streamliners and diesels the railroads began to embrace in the 1930's. Today, the low power consumption of its single light has come to be offset by the high maintenance cost of the movable lens system, so that now trilight signals are preferred for new installations and are replacing searchlights in many places.

A searchlight signal on the Atlantic Coast Line at Pomona, Fla.

## Position-light Signals

Developed by the Pennsylvania Railroad and introduced in 1915, the position-light signal saw use only on the PRR and affiliated lines like Long Island, Norfolk & Western, and Lehigh Valley. In this design, high-intensity lights of a single color (a fog-penetrating yellow) are illuminated in rows of three, forming lines not unlike semaphore signals.

A fully equipped signal head can display lights in horizontal, vertical, and two diagonal rows, but partial units are common, especially in cases where two heads are used. (PRR also had special X and circle aspects.)

While a relatively large number of lights was needed for each unit, the signal could still be read with one light in a row burned out.

High maintenance, and the fact that they do not use the standard red-yellow-green colors, mean that position-lights are now being phased out.

Position-light signals viewed from the fireman's side of a Norfolk & Western steam locomotive.

## Color Position-light Signals

First used in 1921, color position-lights are as identified with Baltimore & Ohio and its related lines (chiefly the old Alton lines) as the position-light is with the PRR. Unlike the position-light, this type has no center lamp and uses larger lenses of three colors. The two lamps in each row are green (vertical), yellow (diagonal), and red (horizontal). The basic design often is modified with lunar white or yellow markers above and below the main head.

Although more costly than some other signal types, a color position-light signal has the advantage of redundancy: it displays its message in two ways: the position of the lights and the color of the lights. As with position-light signals, the color position-light signal can be read with one light in a row burned out.

On the Northeast Corridor, Amtrak is converting position-light signals to color position-lights by removing the center lamp and changing the remaining lenses to color types.

## Cab Signals

An engineman had this view ahead from the cab of a GG1. Visible through the cab window is one of the Pennsylvania's distinctive position light signals, while to the engineman's left are the cab signal indicators and some of the many gauges and meters required for the operation of an electric locomotive.

Some railroads use cab signals (small signals in the locomotive cab) to supplement or even replace wayside signals. These may be associated with an Automatic Train Stop System (ATS), which requires acknowledgment and/or observance of signals.

Cab signals and ATS are yet other types of protective overlays, created to keep trains apart and operations safe.

An engineman had this view ahead from the cab of a GG1. Visible through the cab window is one of the Pennsylvania's distinctive position light signals, while to the engineman's left are the cab signal indicators and some of the many gauges and meters required for the operation of an electric locomotive.

## Rail Traffic Management and Signaling

Train control systems poses high demands on positioning with respect to availability, reliability and integrity. These requirements can only be fulfilled by means of integrated positioning systems, which combine GNSS with other sensors.

The use of GNSS in railway systems presents many advantages, in particular the monitoring of train's exact location, logistic information management, enhanced train signaling (which improves safety, but also enables e.g. reduced distances between trains and therefore increased train frequencies), and the possibility to map the transport infrastructure.

Thus, while the number of applications based on GNSS is considerably behind the number of those used in other domains, such as road transport, incorporating GNSS receivers into modern signaling, train control and other railway systems has become common.

### Application Architecture

Train control, signaling, passenger traffic or transportation of dangerous goods are safety-critical applications, which show very demanding requirements in terms of availability, continuity and integrity. In order to fulfill these high performance demands complementary positioning sensors such as accelerometers or digital track maps and alternative communication components (e.g. GSM-R, UMTS, satcom) have to be grouped around the receiver/communication core.

In railway applications like train control and train supervision a GNSS receiver usually serves as a basic sensor which is enhanced by other navigation sensors and by an integrated fault tolerant sensor fusion software.

Traffic Management and Signaling systems include the Signaling Block systems which encompasses collision avoiding systems. Both the ETCS (European Train Control System) in Europe and the PTC in U.S., deal with train control and signaling systems standardization in order to improve interoperability.

The GNSS receivers for Traffic Management and Signaling are considered safety critical applications.

## European Train Control System

The European Rail Traffic Management System (ERTMS) is the European cross-border interoperable standard for train control, signaling and traffic management. This standard includes two layers that can rely on GNSS information:

1. European Train Control System (ETCS): Is the signaling element of the system which includes the control of movement authorities, automatic train protection and the interface to interlockings.
    - It allows the stepwise reduction of complexity for train drivers (automation of control activities),
    - It brings track side signaling into the driver cabin,
    - It provides information to the onboard display,
    - It allows for permanent train control,
    - Train driver concentrates on core tasks.
2. European Traffic Management Layer (ETML): Is the operation management level intended to optimize train movements by the interpretation of timetables and train running data, involving the improvement of:
    - Real-time train management and route planning,
    - Rail node fluidity,
    - Customer and operating staff information.

The European Train Control System (ETCS) shall replace the national train control systems in the future to ensure cross-border interoperability and improve railway safety and efficient management.

The ETCS deals with control and signaling systems used in remote electronic braking, train diagnostics and train/wagon location monitoring, all of which are functions whose safety and accuracy could especially be improved through application of GNSS. For example, combining GNSS position data with traditional train sensor readings, such as odometry, can improve the ETCS location performance and therefore allow a safe discrimination between parallel tracks and track change at switches.

The availability of satellite signals as well as the geometry of available satellites can be affected by obstacles (e.g. hills, buildings, vegetation, tunnels). Since the GNSS positioning performance also depends on the environmental conditions and the required performance level, positioning availability can not always be achieved due to unavoidable shadowing of GNSS signals under typical train operation environment. Therefore, GNSS location shall not be used as sole means in a train control system. The hybridization with aiding sensors is mandatory and a terrestrial augmentation of SBAS systems may be needed.

## Application Characterization

In general, a GNSS-based train control system will be superior to other train control systems due to.

- Lower cost,
- Better interoperability (no need of infrastructure equipment),
- Performance improvement with respect to the existing odometry.

## ETCS Characterization

The European Train Control System fulfills two main objectives: Define a communication based train control system, which is useful for both high and low speed railway services. Define a standard for a uniform signaling system on an MMI (Man-Machine Interface), so neither locomotives nor drivers have to be replaced when crossing borders. The textual display on the MMI can be made to comply with the driver's language.

The ETCS equipment consists of two main subsystems:

- On-board system, which performs speed supervision and brake intervention,
- Radio Block Centre trackside system, which gives permission to run within a movement authority (MA). The task of this system is to ensure that all trains in the related area are keeping the safety distance to the neighboring trains. Also the supervision of special operational procedures like joining or splitting of trains is a task of this system.

For lines with different capacities three application levels with different types of trackside and on-board equipment are possible.

## PTC Characterization

The PTC systems are comprised of:

- Digital data link communications networks,
- Continuous and accurate positioning systems such as Nationwide Differential Global Positioning System (NDGPS),
- On-board computers with digitized maps on locomotives and maintenance-of-way equipment,
- Throttle-brake interfaces on locomotives,
- Wayside interface units at switches and Wayside detectors,
- Control center computers and Displays.

PTC systems may also interface with tactical and strategic traffic planners, work order reporting systems, and locomotive health reporting systems.

These systems also issue movement authorities to train and maintenance-of-way crews, track the location of the trains and maintenance-of-way vehicles, have the ability to automatically enforce movement authorities, and continually update operating data systems with information on the location of trains, locomotives, cars, and crews.

In addition to providing a greater level of safety and security, PTC systems also enable a railroad to run scheduled operations and provide improved running time, greater running time reliability, higher asset utilization, and greater track capacity. They will assist railroads in measuring and managing costs and in improving energy efficiency.

In the future, GNSS will be integrated in vehicle-to-vehicle communication systems in order to warn trains and cars of potential collisions at railroad crossings.

### Application Examples

In Europe, projects such as GADEROS (Galileo Demonstrator for Railway Operation System), RUNE (Railway User Navigation Equipment), INTEGRAIL and most recently GRAIL were each aimed at supporting the introduction of GNSS in the railway sector (particularly in rail safety applications).

- INTEGRAIL system: The aim of INTEGRAIL system in Europe is to open the way for profitable use of the EGNOS signal in safety-critical railway traffic management and control, achieving significant improvements for the rail traffic operator with respect to cost, redundancy and reliability of the present train speed measurement systems, by adding satellite navigation information and the integrity information offered by EGNOS. The INTEGRAIL system is able to provide reliable position and integrity information under varying operational conditions.
- ECORAIL system: The aim is to design and test a positioning system that is based on satellite navigation in a safety critical railway application.
- GRAIL system: The objective is to support the introduction of GNSS in the Rail market, with a special emphasis on ERTMS/ETCS.

Concerning PTC System Development and Deployments, there are currently 11 different PTC projects in varying stages development and implementation, involving 9 different railroads in at least 16 different States, and consisting of over 4,000 track miles in U.S.

## Signaling Block Systems

Signaling block systems enable the safe and efficient operation of railways by preventing collisions between trains. The basic principle is that a route is broken up into a series of blocks, only one train may occupy a block at a time, and that the blocks are sized to allow a train to stop within them. This ensures that a train always has time to stop before reaching another train on the same line. A block system is referred to as the *method of working* in the UK, *method of operation* in the US and *safeworking* in Australia.

In most examples, a system of signals is used to control flow between the blocks. When a train enters a block, signals at both ends change to indicate that the block is occupied, typically using red lamps or indicator flags. When a train first enters a block, the rear of the same train has not yet left the previous block, so both blocks are marked as occupied. This ensures there is slightly less than one block length on either end of the train that is marked as occupied, so any other train approaching this section will have enough room to stop in time even if the first train stops dead on the tracks. The previously occupied block will only be marked unoccupied when the end of the train has entirely left it, leaving the entire block clear.

Block systems have the disadvantage that they limit the number of trains on a particular route to something smaller than the number of blocks. Since the route has a fixed length, increasing the number of trains requires more blocks, which means the blocks are shorter, which means the trains have to operate at lower speeds in order to safely stop. As a result, the number and size of blocks are strongly related to the route's overall route capacity and cannot be easily changed without changes to the signals all along the line.

Block systems are used to control trains between stations and yards, and not normally within the yards, where some other method will be used. Any block system is defined by its associated physical equipment and by the application of a relevant set of rules. Some systems involve the use of signals while others do not. Some systems are specifically designed for single track railways for which a danger exists of both head-on and rear-end collision, as opposed to double track, whose main danger is a rear-end collision.

## Concepts

The basic problem for train control is the relatively long stopping distances of a loaded train. This is often far longer than the operator's eyesight, especially at night or in bad weather. The distances are great enough that local terrain may block sighting of trains ahead, and even the routing of the rails, around bends and such, may make it difficult to even know where to look for another train.

This leads to the possibility that a train may break down on the tracks, and the following train suddenly comes upon it when rounding a bend, or suddenly sees its rear signal lamp. In these situations there will not be enough room for the train to stop before it collides. This is known as the "brick wall criterion". Even in the case of two fully operational trains, differences in speed may be great enough that a faster train may not have time to slow down to match the speed of the one in front before it overtakes it.

Blocks avoid these problems by ensuring there is a certain minimum distance between trains, a distance that is set to ensure any train operating within the speed and load limits will have time to stop before reaching a train ahead of it. There are many potential solutions to implementing such a system.

## Block Signaling Methods

### Strict Timetable Operation

Most rail routes have a sort of natural block layout inherent in the layout of the railway stations. This provides the ability to implement a set of blocks using manual signaling based at these locations. In this case, the station operator places a flag indicating a train has just left the station, and removes it only after a fixed time.

Trains operate according to a strict timetable, and as such, cannot leave a station until an appointed time, and until any other trains they were to meet at that station have arrived. If one train is delayed, all trains it is scheduled to meet are delayed. This can quickly lead to all trains on the railway being affected.

This method is not authorised for use in many high-traffic railway systems.

## Timetable and Train Order

Popular on single track lines in North America up until the 1980s, Train Order operation was less a block system and more of a system of determining which trains would have the right of way when train movements would come into conflict. Trains would make use of a predetermined operating plan known as the timetable which made use of fixed passing locations often referred to as stations. Amendments to the operating plan would come from a train dispatcher in the form of train orders, transmitted to the trains via intermediaries known as agents or operators at train order stations.

This method is not currently authorised for use in the UK. A similar system, known as *Telegraph and Crossing Order*, was used in the 19th century, but after three serious head-on collisions in the 1870s (Menheniot, Cornwall Railway, 1873; Thorpe, Great Eastern Railway, 1874; Radstock, Somerset & Dorset Railway, 1876) its use was condemned.

In North American train order system was often implemented on top of other block systems when those block systems needed to be superseded. For example, where manual or automatic block was implemented, train orders would be used to authorize movements into occupied blocks, against the current of traffic or where no current of traffic was established.

## One Train Working

## One Train Working (with Train Staff)

If a single track branch line is a dead end with a simple shuttle train service, then a single token is sufficient. The driver of any train entering the branch line (or occupying any part of it) must be in possession of the token, and no collision with another train is possible. For convenience in passing it from hand to hand, the token was often in the form of a staff, typically 800 mm long and 40 mm diameter, and is referred to as a *train staff*. Such a staff may be a wooden staff with a brass plate stating the section of line on which it is valid, or it may be in the form of a key.

In UK terminology, this method of working was originally referred to as *One Engine in Steam (OES).*

## One Train Working (without Train Staff)

A modern variation of the One Train Working system operates without any train staff. On these lines the clearance of the controlled branch entry signal is the driver's sole authority to enter the branch, and once the train has passed that signal, the interlocking will hold it at 'danger' (and the signal cannot be cleared a subsequent time) until the branch service train, on its return journey has sequentially operated two track circuits at the start of the branch. Continuous train detection on the branch is not required. Safety is ensured by the interlocking circuitry, and if a track-circuit failure occurs then special emergency working by pilotman must be introduced.

Accepting a token on the South Devon Railway.

## Token Block

Authority to occupy a block is provided by physical possession of a token by train staff that is obtained in such a way that ensures that only one train is in a section of track at a time.

## Ordinary Train Staff and Ticket (OTST) or (OTS&T)

## Ordinary Train Staff Sections

Some low traffic lines dispensed with Tickets and became Ordinary Train Staff sections (OTS).

### Electric Train Staff (ETS)

These came in two sizes, large and miniature.

## Manual Block System

Authority to occupy a block is conveyed to trains by the use of wayside signals manually controlled by human operators following various procedures to communicate with other block stations to ensure separation of trains.

## Telegraph Block

Used on multiple track sections whereby the passage of trains from one point to the next was controlled by instruments connected by telegraph wires. Used extensively in Australia.

## Telephonic Block

In this system, the occupation of a given section of track between two stations is agreed between its station masters, via telephone. For greater safety there can be additional layers of protection; for example, a regulating post, with supervisory powers connected to all the stations in a line; timetable (Portugal); and/or computer assistance (France).

Portugal, Spain and France still use this system on at least some main lines, although the total length of track governed by this system is decreasing rapidly due to its labour intensity and its inherent perceived lack of safety, relying as it does primarily on human communication (sometimes involving more than just the two station masters at each end of the block) and simple railway interlockings at the stations.

In Portugal, the telephonic block was the main safety system across the national railway network until the mid-1990s due to lack of resources. Thus, it evolved to try to provide multiple layers of safety on busy single-track lines with diverse train types, albeit at the cost of high levels of staffing. In the Portuguese system, although the authority of train movement on the main lines is the sole responsibility of the stations along those lines, a regulating post oversees them and, in case of disagreement, instructs stations as to how the traffic should be organised. On the other hand, each train timetable indicates all interactions with other trains (e.g. crossings with other trains; trains that they overtake; trains that overtake them) clearly marked at the stations at which those interactions should occur. Any deviation from that—arising, for example, from delays or extra trains—must be provided to the train crews in writing. Despite the general practice that, when two trains cross, they both stop at the nearest station, this system allows for good average speeds for fast trains similar to those on an automatic-signaling line. However, if minor delays occur and then proliferate, longer delays can arise as the system's additional safety mode is invoked (i.e. the paperwork-intensive process of updating train-movement instructions to reflect the altered crossing patterns). Such delays would not happen, at least not for the same reason, on an automatic-signaling line.

In general, the system dictates that a block is assumed to be closed; that is, permission must be obtained before a train is allowed to enter a block at one station en route to the other. However, in France, on multiple tracks, the block is usually open in unidirectional track sections. That is, after a station confirms that a previous train has vacated the block, the next train travelling in the same direction can immediately enter the block, with the station master at the entry station informing the exit station of the time that the train entered the block.

## Tokenless Block

This a system for use on single track railways, which requires neither the use of tokens nor provision of continuous train detection through the section. The signaling is designed such that the controlling signals will only allow one train to enter the line. The signalman at the far end of the section must visually check that the whole train has left the section and not become divided.

## Automatic Block Signaling

Vertical colour light signal on the Enshū Railway Line.

Automatic block signaling uses a series of automated signals, normally lights or flags, that change their display, or *aspect*, based on the movement of trains past a sensor. This is by far the most common type of block system As of 2018, used in almost every type of railway from rapid transit systems to railway mainlines. There are a wide variety of systems, and an even wider variety of signals, but they all work in roughly the same fashion.

Like the manual block systems outlined above, automatic systems divide the route into fixed blocks. At the end of each block, a set of signals is installed, along with a track-side sensor. When a train passes the sensor, the signals are triggered to display the "block occupied" aspect on the signals at either end of that block. In most systems the signals do not immediately return to the "block empty" aspect when the train leaves, instead there is some sort of mechanical delay that retains the block occupied aspect, or more commonly, presents a "proceed with caution" aspect.

### Moving Blocks

In terms of ensuring safety, the real consideration is the stopping distance of a given train and the distance at which it can spot another train. Blocks do not actually implement this concept, they implement a signaling system that ensures the worst performing train on a line has enough time to stop. This means any train with better stopping performance is forced to operate at speeds that are lower than its maximum, unless all of the trains on a particular line are identical.

The key issue is that a given train cannot safely see another train in time to stop. However, this is not true for trains that are equipped with some sort of inter-train communications system. In this case, any given train can keep itself at a safe distance from other trains, without the need for fixed blocks. These moving block systems have become popular since the required technology first started appearing in the 1970s.

In such systems, any train on the route can listen for signals from all the other trains, and then move in a way to ensure they have enough distance to stop. Early moving block systems used a cable strung along the rail line. Trains would use magnetic inductance to inject signals into the line indicating their location. The cable could also provide that location in a variety of ways that could be picked up by a sensor on the train. More modern systems may use off-board location systems like Global Positioning System or track-side indicators, and send the data between the trains using various radio-based methods.

The advantage to moving block systems is that there is no fixed number of trains on the line, because there are no fixed blocks. This can greatly improve route capacity, as seen in the Jubilee line and Northern line on the London Underground, where upgrades for the 2012 Summer Olympics improved capacity by about 50%.

## Cab Signaling

Cab signaling is a railway safety system that communicates track status and condition information to the cab, crew compartment or driver's compartment of a locomotive, railcar or multiple unit. The information is continually updated giving an easy to read display to the train driver or engine driver.

The simplest systems display the trackside signal, while more sophisticated systems also display allowable speed, location of nearby trains, and dynamic information about the track ahead. Cab signals can also be part of a more comprehensive train protection system that can automatically apply the brakes stopping the train if the operator does not respond appropriately to a dangerous condition.

The main purpose of a signal system is to enforce a safe separation between trains and to stop or slow trains in advance of a restrictive situation. The cab signal system is an improvement over the wayside signal system, where visual signals beside or above the right-of-way govern the movement of trains, as it provides the train operator with a continuous reminder of the last wayside signal or a continuous indication of the state of the track ahead.

The first such systems were installed on an experimental basis in the 1910s in the United Kingdom, in the 1920s in the United States, and in the Netherlands in the 1940s. Modern high-speed rail systems such as those in Japan, France, and Germany were all designed from the start to use in-cab signaling due to the impracticality of sighting wayside signals at the new higher train speeds. Worldwide, legacy rail lines continue to see limited adoption of Cab Signaling outside of high density or suburban rail districts and in many cases is precluded by use of older intermittent Automatic Train Stop technology.

In North America, the coded track circuit system developed by the Pennsylvania Railroad (PRR) and Union Switch & Signal (US&S) became the de facto national standard. Variations of this system are also in use on many rapid transit systems and form the basis for several international cab signaling systems such as CAWS in Ireland, BACC in Italy, ALSN in Russia and the first generation Shinkansen signaling developed by Japan National Railways (JNR).

In Europe and elsewhere in the world, cab signaling standards were developed on a country by country basis with limited interoperability, however new technologies like the European Rail Traffic Management System (ERTMS) aim to improve interoperability. The train-control component of ERTMS, termed European Train Control System (ETCS), is a functional specification that incorporates some of the former national standards and allows them to be fully interoperable with a few modifications.

## Cab Signal Types

All cab signaling systems must have a continuous in-cab indication to inform the driver of track condition ahead; however, these fall into two main categories. Intermittent cab signals are updated at discrete points along the rail line and between these points the display will reflect information from the last update. Continuous cab signals receive a continuous flow of information about the state of the track ahead and can have the cab indication change at any time to reflect any updates. The majority of cab signaling systems, including those that use coded track circuits, are continuous.

### Intermittent

The German Indusi and Dutch ATB-NG fall into this category. These and other such systems provide constant reminders to drivers of track conditions ahead, but are only updated at discrete points. This can lead to situations where the information displayed to the driver has become out of date. Intermittent cab signaling systems have functional overlap with many other train protection systems such as trip stops, but the distinction is that a driver or automatic operating system makes continuous reference to the last received update.

## Continuous

Continuous systems have the added benefit of fail safe behaviour in the event a train stops receiving the continuous event relied upon by the cab signaling system. Early systems use the rails or loop conductors laid along the track to provide continuous communication between wayside signal systems and the train. These systems provided for the transmission of more information than was typically possible with contemporary intermittent systems and are what enabled the ability to display a miniature signal to the driver; hence the term, "cab signaling". Continuous systems are also more easily paired with Automatic Train Control technology, which can enforce speed restrictions based on information received through the signaling system, because continuous cab signals can change at any time to be more or less restrictive, providing for more efficient operation than intermittent ATC systems.

## Information Transmission

Cab signals require a means of transmitting information from wayside to train. There are a few main methods to accomplish this information transfer.

## Electric or Magnetic

This is popular for early intermittent systems that used the presence of a magnetic field or electric current to designate a hazardous condition. The British Rail Automatic Warning System (AWS) is an example of a two-indication cab signal system transmitting information using a magnetic field.

## Inductive

Inductive systems are non-contact systems that rely on more than the simple presence or absence of a magnetic field to transmit a message. Inductive systems typically require a beacon or an induction loop to be installed at every signal and other intermediate locations. The inductive coil uses a changing magnetic field to transmit messages to the train. Typically, the frequency of pulses in the inductive coil are assigned different meanings. Continuous inductive systems can be made by using the running rails as one long tuned inductive loop.

Examples of intermittent inductive systems include the German Indusi system. Continuous inductive systems include the two-aspect: General Railway Signal Company "Automatic Train Control" installed on the Chicago and North Western Railroad among others.

## Coded Track Circuits

A coded track circuit based system is essentially an inductive system that uses the running rails as information transmitter. The coded track circuits serve a dual purpose: to perform the train detection and rail continuity detection functions of a standard track circuit, and to continuously transmit signal indications to the train. The coded track circuit systems eliminate the need for specialized beacons.

Examples of coded track circuit systems include the Pennsylvania Railroad standard system, a variation of which was used on the London Underground Victoria line, Later, audio frequency (AF) track circuit systems eventually came to replace "power" frequency systems in rapid transit applications as higher frequency signals could self-attenuate reducing the need for insulated rail joints. Some of the first users of AF cab signal systems include the Washington Metro and Bay Area Rapid Transit. More recently, digital systems have become preferred, transmitting speed information to trains using datagrams instead of simple codes. The French TVM makes use of the running rails to transmit the digital signaling information, while the German LZB system makes use of auxiliary wires strung down the centre of the track to continually transmit the signaling information.

## Transponder

Transponder based systems make use of fixed antenna loops or beacons (called balises) that transmit datagrams or other information to a train as it passes overhead. While similar to intermittent inductive systems, transponder based cab signaling transmit more information and can also receive information from the train to aid traffic management. The low cost of loops and beacons allows for a larger number of information points that may have been possible with older systems as well as finer grained signaling information. The British Automatic Train Protection was one example of this technology along with the more recent Dutch ATB-NG.

## Wireless

Wireless cab signaling systems dispense with all track-based communications infrastructure and instead rely on fixed wireless transmitters to send trains signaling information. This method is most closely associated with communications-based train control. ETCS levels 2 and 3 make use of this system, as do a number of other cab signaling systems under development.

## Cab Display Unit

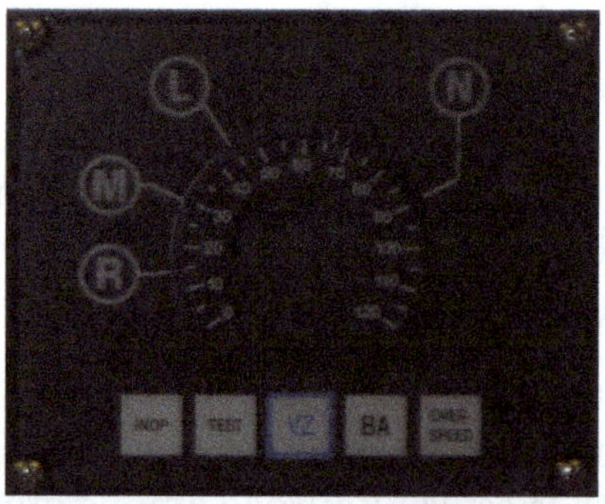


CDU used on Metro-North is integrated with the speedometer indicating which speed limit applies for each signal.

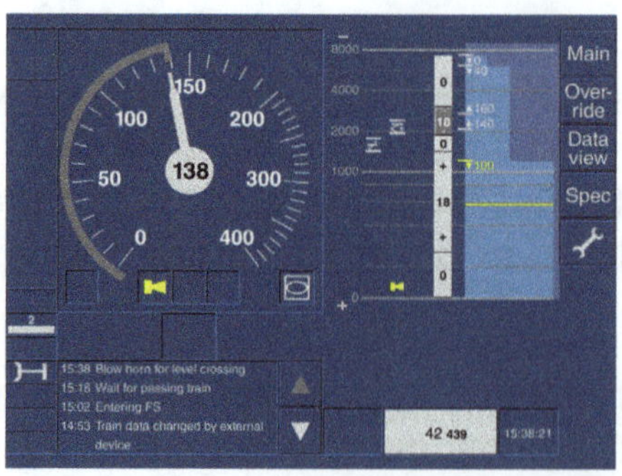


The ETCS driver machine interface.

The cab display unit (CDU), (also called a driver machine interface (DMI) in the ERTMS standard) is the interface between the train operator and the cab signaling system. Early CDU's displayed simple warning indications or representations of wayside railway signals. Later, many railways and rapid transit systems would dispense with miniature in-cab signals in favour of an indication of what speed the operator was permitted to travel at. Typically this was in conjunction with some sort of Automatic Train Control speed enforcement system where it becomes more important for operators to run their trains at specific speeds instead of using their judgement based on signal indications. One common innovation was to integrate the speedometer and cab signal display, superimposing or juxtaposing the allowed speed with the current speed. Digital cab signaling systems that make use of datagrams with "distance to target" information can use simple displays that simply inform the driver when they are approaching a speed penalty or have triggered a speed penalty or more complex ones that show a moving graph of the minimum braking curves permitted to reach the speed target.

CDU's also inform the operator which, if any, mode the system might be in or if it is active at all. CDU's can also be integrated into the alertness system, providing count-downs to the alertness penalty or a means by which to cancel the alarm.

## Cab Signaling Systems in the United States

Cab signaling in the United States was driven by a 1922 ruling by the Interstate Commerce Commission (ICC) that required 49 railways to install some form of automatic train control in one full passenger division by 1925. While several large railways, including the Santa Fe and New York Central, fulfilled the requirement by installing intermittent inductive train stop devices, the PRR saw an opportunity to improve operational efficiency and installed the first continuous cab signal systems, eventually settling on pulse code cab signaling technology supplied by Union Switch and Signal.

In response to the PRR lead, the ICC mandated that some of the nation's other large railways must equip at least one division with continuous cab signal technology as a test to compare technologies and operating practices. The affected railroads were less than enthusiastic, and many chose to equip one of their more isolated or less trafficked routes to minimize the number of locomotives to be equipped with the apparatus.

Amtrak ACSES-capable cab signal display unit showing both a miniature signal and associated speed limit.

Several railways chose the inductive loop system rejected by the PRR. These railways included the Central Railroad of New Jersey (installed on its Southern Division), the Reading Railroad (installed on its Atlantic City Railroad main line) and the New York Central. Both the Chicago Northwestern and Illinois Central employed a two-aspect system on select suburban lines near Chicago. The cab signals would display "Clear" or "Restricting" aspects. The CNW went further and eliminated the wayside intermediate signals in the stretch of track between Elmhurst and West Chicago, requiring trains to proceed solely based on the 2-aspect cab signals. The Chicago, Milwaukee, St. Paul and Pacific Railroad had a 3-aspect system operating by 1935 between Portage, Wisconsin and Minneapolis, Minnesota.

As the Pennsylvania Railroad system was the only one adopted on a large scale, it became a de facto national standard, and most installations of cab signals in the current era have been this type. Recently, there have been several new types of cab signaling which use communications-based technology to reduce the cost of wayside equipment or supplement existing signal technologies to enforce speed restrictions and absolute stops and to respond to grade crossing malfunctions or incursions.

The first of these was the Speed Enforcement System (SES) employed by New Jersey Transit on their low-density Pascack Valley Line as a pilot program using a dedicated fleet of 13 GP40PH-2 locomotives. SES used a system of transponder beacons attached to wayside block signals to enforce signal speed. SES was disliked by engine crews due to its habit of causing immediate penalty brake applications without first sounding an overspeed alarm and giving the engineer a chance to decelerate. SES is in the process of being removed from this line, and is being replaced with CSS.

Amtrak uses the Advanced Civil Speed Enforcement System (ACSES) for its *Acela Express* high-speed rail service on the NEC. ACSES was an overlay to the existing PRR-type CSS and uses the same SES transponder technology to enforce both permanent and temporary speed restrictions at curves and other geographic features. The on-board cab signal unit processes both the pulse code "signal speed" and the ACSES "civil speed", then enforces the lower of the two. ACSES also provides for a positive stop at absolute signals which could be released by a code provided by the dispatcher transmitted from the stopped locomotive via a data radio. Later this was amended to a simpler "stop release" button on the cab signal display.

## Communications-based Train Control

Communications-based train control (CBTC) is a railway signaling system that makes use of the telecommunications between the train and track equipment for the traffic management and infrastructure control. By means of the CBTC systems, the exact position of a train is known more accurately than with the traditional signaling systems. This results in a more efficient and safe way to manage the railway traffic. Metros (and other railway systems) are able to improve headways while maintaining or even improving safety.

A CBTC system is a "continuous, automatic train control system utilizing high-resolution train location determination, independent from track circuits; continuous, high-capacity, bidirectional train-to-wayside data communications; and trainborne and wayside processors capable of implementing automatic train protection (ATP) functions, as well as optional automatic train operation (ATO) and automatic train supervision (ATS) functions," as defined in the IEEE 1474 standard.

The main objective of CBTC is to increase capacity by reducing the time interval (headway) between trains.

Traditional signaling systems detect trains in discrete sections of the track called 'blocks', each protected by signals that prevent a train entering an occupied block. Since every block is a fixed section of track, these systems are referred to as fixed block systems.

In a moving block CBTC system the protected section for each train is a "block" that moves with and trails behind it, and provides continuous communication of the train's exact position via radio, inductive loop, etc.

SFO AirTrain, in San Francisco Airport, was the first radio-based CBTC system.

As a result, Bombardier opened the world's first radio-based CBTC system at San Francisco airport's automated people mover (APM) in February 2003. A few months later, in June 2003, Alstom introduced the railway application of its radio technology on the Singapore North East Line. Previously, CBTC has its former origins in the loop based systems developed by Alcatel SEL (now Thales) for the Bombardier Automated Rapid Transit (ART) systems in Canada during the mid-1980s. These systems, which were also referred to as transmission-based train control (TBTC), made use of inductive loop transmission techniques for track to train communication, introducing an alternative to track circuit based communication. This technology, operating in the 30–60 kHz frequency range to communicate trains and wayside equipment, was widely adopted by the metro operators in spite of some electromagnetic compatibility (EMC) issues, as well as other installation and maintenance concerns.

As with new application of any technology, some problems arose at the beginning mainly due to compatibility and interoperability aspects. However, there have been relevant improvements since then, and currently the reliability of the radio-based communication systems has grown significantly.

Systems using radio communication technology are considered to be CBTC systems. So, for clarity and to keep in line with the state-of-the-art solutions for operator's requirements, this article only covers the latest moving block principle based (either true moving block or virtual block, so not dependent on track-based detection of the trains).

## Main Features

### CBTC and Moving Block

CBTC systems are modern railway signaling systems that can mainly be used in urban railway lines (either light or heavy) and APMs, although it could also be deployed on commuter lines.

For main lines, a similar system might be the European Railway Traffic Management System ERTMS Level 3 (not yet fully defined). In the modern CBTC systems the trains continuously calculate and communicate their status via radio to the wayside equipment distributed along the line. This status includes, among other parameters, the exact position, speed, travel direction and braking distance. This information allows calculation of the area potentially occupied by the train on the track. It also enables the wayside equipment to define the points on the line that must never be passed by the other trains on the same track. These points are communicated to make the trains automatically and continuously adjust their speed while maintaining the safety and comfort (jerk) requirements. So, the trains continuously receive information regarding the distance to the preceding train and are then able to adjust their safety distance accordingly.

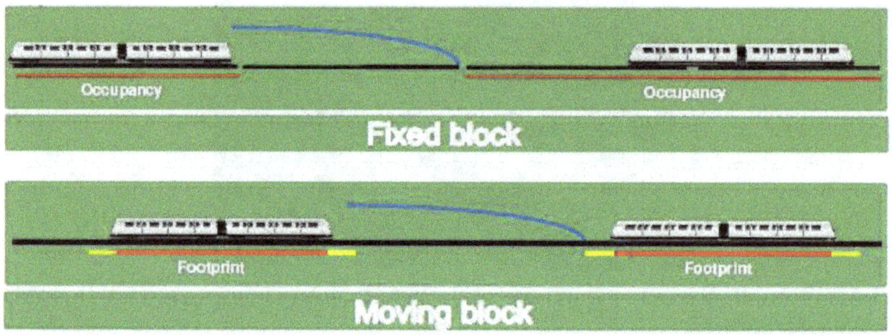


Safety distance (safe-braking distance) between trains in fixed block and moving block signaling systems.

From the signaling system perspective, the first figure shows the total occupancy of the leading train by including the whole blocks which the train is located on. This is due to the fact that it is impossible for the system to know exactly where the train actually is within these blocks. Therefore, the fixed block system only allows the following train to move up to the last unoccupied block's border.

In a moving block system as shown in the second figure, the train position and its braking curve is continuously calculated by the trains, and then communicated via radio to the wayside equipment. Thus, the wayside equipment is able to establish protected areas, each one called Limit of Movement Authority (LMA), up to the nearest obstacle (in the figure the tail of the train in front). Movement Authority (MA) is the permission for a train to move to a specific location within the constraints of the infrastructure and with supervision of speed. End of Authority is the location to which the train is permitted to proceed and where target speed is equal to zero. End of Movement is the location to which the train is permitted to proceed according to an MA. When transmitting a MA, it is the end of the last section given in the MA.

It is important to mention that the occupancy calculated in these systems must include a safety margin for location uncertainty (in yellow in the figure) added to the length of the train. Both of them form what is usually called 'Footprint'. This safety margin depends on the accuracy of the odometry system in the train.

CBTC systems based on moving block allows the reduction of the safety distance between two consecutive trains. This distance is varying according to the continuous updates of the train location and speed, maintaining the safety requirements. This results in a reduced headway between consecutive trains and an increased transport capacity.

## Grades of Automation

Modern CBTC systems allow different levels of automation or *Grades of Automation* (GoA), as defined and classified in the IEC 62290-1. In fact, CBTC is not a synonym for "driverless" or "automated trains" although it is considered as a basic enabler technology for this purpose.

The grades of automation available range from a manual protected operation, GoA 1 (usually applied as a fallback operation mode) to the fully automated operation, GoA 4 (Unattended Train Operation, UTO). Intermediate operation modes comprise semi-automated GoA 2 (Semi-automated Operation Mode, STO) or driverless GoA 3 (Driverless Train Operation, DTO). The latter operates without a driver in the cabin, but requires an attendant to face degraded modes of operation as well as guide the passengers in the case of emergencies. The higher the GoA, the higher the safety, functionality and performance levels must be.

## Main Applications

CBTC systems allow optimal use of the railway infrastructure as well as achieving maximum capacity and minimum headway between operating trains, while maintaining the safety requirements. These systems are suitable for the new highly demanding urban lines, but also to be overlaid on existing lines in order to improve their performance.

Of course, in the case of upgrading existing lines the design, installation, test and commissioning stages are much more critical. This is mainly due to the challenge of deploying the overlying system without disrupting the revenue service.

## Main Benefits

The evolution of the technology and the experience gained in operation over the last 30 years means that modern CBTC systems are more reliable and less prone to failure than older train control systems. CBTC systems normally have less wayside equipment and their diagnostic and monitoring tools have been improved, which makes them easier to implement and, more importantly, easier to maintain.

CBTC technology is evolving, making use of the latest techniques and components to offer more compact systems and simpler architectures. For instance, with the advent of modern electronics it has been possible to build in redundancy so that single failures do not adversely impact operational availability.

Moreover, these systems offer complete flexibility in terms of operational schedules or timetables, enabling urban rail operators to respond to the specific traffic demand more swiftly and efficiently and to solve traffic congestion problems. In fact, automatic operation systems have the potential to significantly reduce the headway and improve the traffic capacity compared to manual driving systems.

Finally, it is important to mention that the CBTC systems have proven to be more energy efficient than traditional manually driven systems. The use of new functionalities, such as automatic driving strategies or a better adaptation of the transport offer to the actual demand, allows significant energy savings reducing the power consumption.

## Risks

The primary risk of an electronic train control system is that if the communications link between any of the trains is disrupted then all or part of the system might have to enter a failsafe state until the problem is remedied. Depending on the severity of the communication loss, this state can range from vehicles temporarily reducing speed, coming to a halt or operating in a degraded mode until communications are re-established. If communication outage is permanent some sort of contingency operation must be implemented which may consist of manual operation using absolute block or, in the worst case, the substitution of an alternative form of transportation. As a result, high availability of CBTC systems is crucial for proper operation, especially if such systems are used to increase transport capacity and reduce headway. System redundancy and recovery mechanisms must then be thoroughly checked to achieve a high robustness in operation. With the increased availability of the CBTC system, there is also a need for extensive training and periodical refresh of system operators on the recovery procedures. In fact, one of the major system hazards in CBTC systems is the probability of human error and improper application of recovery procedures if the system becomes unavailable.

Communications failures can result from equipment malfunction, electromagnetic interference, weak signal strength or saturation of the communications medium. In this case, an interruption can result in a service brake or emergency brake application as real time situational awareness is a critical safety requirement for CBTC and if these interruptions are frequent enough it could seriously impact service. This is the reason why, historically, CBTC systems first implemented radio communication systems in 2003, when the required technology was mature enough for critical applications.

In systems with poor line of sight or spectrum/bandwidth limitations a larger than anticipated number of transponders may be required to enhance the service. This is usually more of an issue with applying CBTC to existing transit systems in tunnels that were not designed from the outset to support it. An alternate method to improve system availability in tunnels is the use of leaky feeder cable that, while having higher initial costs (material + installation) achieves a more reliable radio link.

With the emerging services over open ISM radio bands (i.e. 2.4 GHz and 5.8 GHz) and the potential disruption over critical CBTC services, there is an increasing pressure in the international community (ref. report 676 of UITP organization, Reservation of a Frequency Spectrum for Critical Safety Applications dedicated to Urban Rail Systems) to reserve a frequency band specifically for radio-based urban rail systems. Such decision would help standardize CBTC systems across the market (a growing demand from most operators) and ensure availability for those critical systems.

As a CBTC system is required to have high availability and particularly, allow for a graceful degradation, a secondary method of signaling might be provided to ensure some level of non-degraded service upon partial or complete CBTC unavailability. This is particularly relevant for brownfield implementations (lines with an already existing signaling system) where the infrastructure design cannot be controlled and coexistence with legacy systems is required, at least, temporarily. For example, the New York City Canarsie Line was outfitted with a backup automatic block signaling system capable of supporting 12 trains per hours (tph), compared with the 26 tph of the CBTC system. Although this is a rather common architecture for resignaling projects, it can negate some of the cost savings of CBTC if applied to new lines. This is still a key point in the CBTC development (and is still being discussed), since some providers and operators argue that a fully redundant architecture of the CBTC system may however achieve high availability values by itself.

In principle, CBTC systems may be designed with centralized supervision systems in order to improve maintainability and reduce installation costs. If so, there is an increased risk of a single point of failure that could disrupt service over an entire system or line. Fixed block systems usually work with distributed logic that are normally more resistant to such outages. Therefore, a careful analysis of the benefits and risks of a given CBTC architecture (centralized vs. distributed) must be done during system design.

When CBTC is applied to systems that previously ran under complete human control with operators working on sight it may actually result in a reduction in capacity (albeit with an increase in safety). This is because CBTC operates with less positional certainty than human sight and also with greater margins for error as worst-case train parameters are applied for the design (e.g. guaranteed emergency brake rate vs. nominal brake rate). For instance, CBTC introduction in Philly's Center City trolley tunnel resulted initially in a marked increase in travel time and corresponding decrease in capacity when compared with the unprotected manual driving. This was the offset to finally eradicate vehicle collisions which on-sight driving cannot avoid and showcases the usual conflicts between operation and safety.

## Architecture

The typical architecture of a modern CBTC system comprises the following main subsystems:

- Wayside equipment, which includes the interlocking and the subsystems controlling every zone in the line or network (typically containing the wayside ATP and ATO functionalities). Depending on the suppliers, the architectures may be centralized or distributed. The control of the system is performed from a central command ATS, though local control subsystems may be also included as a fallback.
- CBTC onboard equipment, including ATP and ATO subsystems in the vehicles.
- Train to wayside communication subsystem, currently based on radio links.

Thus, although a CBTC architecture is always depending on the supplier and its technical approach, the following logical components may be found generally in a typical CBTC architecture:

- Onboard ETCS system: This subsystem is in charge of the continuous control of the train speed according to the safety profile, and applying the brake if it is necessary. It is also in charge of the communication with the wayside ATP subsystem in order to exchange the information needed for a safe operation (sending speed and braking distance, and receiving the limit of movement authority for a safe operation).
- Onboard ATO system: It is responsible for the automatic control of the traction and braking effort in order to keep the train under the threshold established by the ATP subsystem. Its main task is either to facilitate the driver or attendant functions, or even to operate the train in a fully automatic mode while maintaining the traffic regulation targets and passenger comfort. It also allows the selection of different automatic driving strategies to adapt the runtime or even reduce the power consumption.

- Wayside ETCS system: This subsystem undertakes the management of all the communications with the trains in its area. Additionally, it calculates the limits of movement authority that every train must respect while operating in the mentioned area. This task is therefore critical for the operation safety.
- Wayside ATO system: It is in charge of controlling the destination and regulation targets of every train. The wayside ATO functionality provides all the trains in the system with their destination as well as with other data such as the dwell time in the stations. Additionally, it may also perform auxiliary and non-safety related tasks including for instance alarm/event communication and management, or handling skip/hold station commands.
- Communication system: The CBTC systems integrate a digital networked radio system by means of antennas or leaky feeder cable for the bi-directional communication between the track equipment and the trains. The 2,4 GHz band is commonly used in these systems (same as WiFi), though other alternative frequencies such as 900 MHz (US), 5.8 GHz or other licensed bands may be used as well.
- ATS system: The ATS system is commonly integrated within most of the CBTC solutions. Its main task is to act as the interface between the operator and the system, managing the traffic according to the specific regulation criteria. Other tasks may include the event and alarm management as well as acting as the interface with external systems.
- Interlocking system: When needed as an independent subsystem (for instance as a fallback system), it will be in charge of the vital control of the trackside objects such as switches or signals, as well as other related functionality. In the case of simpler networks or lines, the functionality of the interlocking may be integrated into the wayside ATP system.

## Centralized Traffic Control

Centralized traffic control (CTC) is a form of railway signaling that originated in North America. CTC consolidates train routing decisions that were previously carried out by local signal operators or the train crews themselves. The system consists of a centralized train dispatcher's office that controls railroad interlockings and traffic flows in portions of the rail system designated as CTC territory. One hallmark of CTC is a control panel with a graphical depiction of the railroad. On this panel, the dispatcher can keep track of trains' locations across the territory that the dispatcher controls. Larger railroads may have multiple dispatcher's offices and even multiple dispatchers for each operating division. These offices are usually located near the busiest yards or stations, and their operational qualities can be compared to air traffic towers.

Key to the concept of CTC is the notion of *traffic control* as it applies to railroads. Trains moving in opposite directions on the same track cannot pass each other without special infrastructure such as sidings and switches that allow one of the trains to move out of the way. Initially, the only two ways for trains to arrange such interactions was to somehow arrange it in advance or provide a communications link between the authority for train movements (the dispatcher) and the trains themselves. These two mechanisms for control would be formalized by railroad companies in a set of procedures called train order operation, which was later partly automated through use of Automatic Block Signals (ABS).

The starting point of each system was the railroad timetable that would form the advanced routing plan for train movements. Trains following the timetable would know when to take sidings, switch tracks and which route to take at junctions. However, if train movements did not go as planned, the timetable would then fail to represent reality, and attempting to follow the printed schedule could lead to routing errors or even accidents. This was especially common on single-track lines that comprised the majority of railroad route miles in North America. Pre-defined "meets" could lead to large delays if either train failed to show up, or worse, an "extra" train not listed in the timetable could suffer a head-on collision with another train that did not expect it.

Therefore, timetable operation was supplemented with train orders, which superseded the instructions in the timetable. From the 1850s until the middle of the twentieth century, train orders were telegraphed in Morse code by a dispatcher to a local station, where the orders would be written down on standardized forms and a copy provided to the train crew when they passed that station, directing them to take certain actions at various points ahead: for example, take a siding to meet another train, wait at a specified location for further instructions, run later than scheduled, or numerous other actions. The development of Direct Traffic Control via radio or telephone between dispatchers and train crews made telegraph orders largely obsolete by the 1970s.

Where traffic density warranted it, multiple tracks could be provided, each with a timetable-defined flow of traffic which would eliminate the need for frequent single track-style "meets." Trains running counter to this flow of traffic would still require train orders, but other trains would not. This system was further automated by the use of Automatic Block Signaling and interlocking towers which allowed for efficient and failsafe setting of conflicting routes at junctions and that kept trains following one another safely separated. However, any track that supported trains running bi-directionally, even under ABS protection, would require further protection to avoid the situation of two trains approaching each other on the same section of track. Such a scenario not only represents a safety hazard, but also would require one train to reverse direction to the nearest passing point.

Before the advent of CTC there were a number of solutions to this problem that did not require the construction of multiple single direction tracks. Many western railroads used an automatic system called absolute permissive block (APB), where trains entering a stretch of single track would cause all of the opposing signals between there and the next passing point to "tumble down" to a Stop position thus preventing opposing trains from entering. In areas of higher traffic density, sometimes bi-directional operation would be established between manned interlocking towers. Each section of bi-directional track would have a traffic control lever associated with it to establish the direction of traffic on that track. Often, both towers would need to set their traffic levers in the same way before a direction of travel could be established. Block signals in the direction of travel would display according to track conditions and signals against the flow of traffic would always be set to their most restrictive aspect. Furthermore, no train could be routed into a section of track against its flow of traffic and the traffic levers would not be able to be changed until the track section was clear of trains. Both APB and manual traffic control would still require train orders in certain situations, and both required trade-offs between human operators and granularity of routing control.

## Development and Technology

CTC panel at the Promenade Street Tower, Providence, Rhode Island. US&S installed the equipment in 1946 for the New York, New Haven and Hartford Railroad.

Penn Central Southern Region (Columbus Division) Train Dispatcher controlling train movements at the CTC "B" board in Columbus, Ohio. At this position, one person could handle about 25 through train movements a day.

The ultimate solution of the costly and imprecise train order system was developed by the General Railway Signal company as their trademarked "Centralized Traffic Control" technology. Its first installation in 1927 was on a 40-mile stretch of the New York Central Railroad between Stanley and Berwick, Ohio, with the CTC control machine located at Fostoria, Ohio. CTC was designed to enable the train dispatcher to control train movements directly, bypassing local operators and eliminating written train orders. Instead, the train dispatcher could directly see the trains' locations and efficiently control the train's movements by displaying signals and controlling switches. It was also designed to enhance safety by reporting any track occupancy to a human operator and automatically preventing trains from entering a track against the established flow of traffic.

What made CTC machines different from standard interlocking machines and ABS was that the vital interlocking hardware was located at the remote location and the CTC machine only displayed track state and sent commands to the remote locations. A command to display a signal would require the remote interlocking to set the flow of traffic and check for a clear route through the interlocking. If a command could not be carried out due to the interlocking logic, the display would not change on the CTC machine. This system provided the same degree flexibility that the manual traffic control has before it, but without the cost and complexity associated with providing a manned operator at the end of every route segment. This was especially true for lightly used lines that could never hope to justify so much overhead.

Initially the communication was accomplished by dedicated wires or wire pairs, but later this was supplanted by pulse code systems utilizing a single common communications link and relay-based telecommunications technology similar to that used in crossbar switches. Also, instead of only displaying information about trains approaching and passing through interlockings, the CTC machine displayed the status of every block between interlockings, where previously such sections had been considered "dark territory" (i.e., of unknown status) as far as the dispatcher was concerned. The CTC system would allow the flow of traffic to be set over many sections of track by a single person at a single location as well as control of switches and signals at interlockings, which also came to be referred to as control points.

CTC machines started out as small consoles in existing towers only operating a few nearby remote interlockings and then grew to control more and more territory, allowing less trafficked towers to be closed. Over time, the machines were moved directly into dispatcher offices, eliminating the need for dispatchers to first communicate with block operators as middlemen. In the late 20th century, the electromechanical control and display systems were replaced with computer operated displays. While similar signaling control mechanisms have been developed in other countries, what sets CTC apart is the paradigm of independent train movement between fixed points under the control and supervision of a central authority.

## Signals and Controlled Points

Newly installed CTC automatic block signals along the Union Pacific Railroad Yuma Subdivision, Coachella.

CTC makes use of railway signals to convey the dispatcher's instructions to the trains. These take the form of routing decisions at controlled points that authorize a train to proceed or stop. Local signaling logic will ultimately determine the exact signal to display based on track occupancy status ahead and the exact route the train needs to take, so the only input required from the CTC system amounts to the go, no-go instruction.

Signals in CTC territory are one of two types: an absolute signal, which is directly controlled by the train dispatcher and helps design the limits of a control point, or an intermediate signal, which is automatically controlled by the conditions of the track in that signal's block and by the condition of the following signal. Train dispatchers cannot directly control intermediate signals and so are almost always excluded from the dispatcher's control display except as an inert reference.

The majority of control points are equipped with remote control, power-operated switches. These switches often are *dual-controlled switches*, as they may be either remotely controlled by the train dispatcher or by manually operating a lever or pump on the switch mechanism itself (although the train dispatcher's permission is generally required to do so). These switches may lead to a passing siding, or they may take the form of a crossover, which allows movement to an adjacent track, or a "turnout" which routes a train to an alternate track (or route).

## Operation

Although some railroads still rely on older, simpler electronic lighted displays and manual controls, in modern implementations, dispatchers rely on computerized systems similar to supervisory control and data acquisition (SCADA) systems to view the location of trains and the aspect, or display, of absolute signals. Typically, these control machines will prevent the dispatcher from giving two trains conflicting authority without needing to first have the command fail at the remote interlocking. Modern computer systems generally display a highly simplified mock-up of the track, displaying the locations of absolute signals and sidings. Track occupancy is displayed via bold or colored lines overlaying the track display, along with tags to identify the train (usually the number of the lead locomotive). Signals which the dispatcher can control are represented as either at stop (typically red) or "displayed" (typically green). A displayed signal is one which is not displaying Stop and the exact aspect that the crew sees is not reported to the dispatcher.

Computer-based controls for a modern electronic interlocking.

## Direct Traffic Control

Direct traffic control (DTC) is a system for authorizing track occupancy used on some railroads instead of or in addition to signals. It is known as "direct" traffic control because the train dispatcher gives track authority directly to the train crew via radio, as opposed to through wayside personnel via telephone or telegraph, as in train orders.

### Layout

A DTC block sign for the Luthman block on the Alaska Railroad.

In DTC, controlled tracks and sidings (those requiring authority from the train dispatcher to occupy) are divided into pre-specified blocks. In addition to being listed by milepost in the railroad's timetable, block limits are delineated by conspicuous signs along the tracks. Every portion of controlled track belongs to a block, as blocks are laid out back-to-back along the entire length of the rail line.

For example, a 30-mile (48 km) length of main line track may be divided into three blocks, Anna, Bess, and Cloy, each 10 miles (16 km) long. At milepost 10, there will be a sign displaying the end of Anna block and the beginning of Bess block, and a similar sign between Bess and Cloy. A train authorized in Anna through Bess blocks must stop before Bess block ends and Cloy block begins.

For example, a length of track may consist of three blocks. Anna, Bess, and Cloy. There is a siding at Bess, and this siding is specified as Bess Siding. Both Bess and Bess Siding blocks extend to the switches at either end of the siding. Anna and Cloy blocks begin on the other side of their respective switches. Neither Bess nor Bess Siding blocks may be occupied without authority from the train dispatcher.

In this example, a train may have authority straight through from Anna through Cloy. Or, it could have authority in Anna through Bess Siding, in which case it must throw the switch between Anna and Bess and enter the siding—this will usually be done to clear up the main track in preparation for an opposing train to pass. After clearing its authority in Anna siding and the opposing train passes, it might then get authority from Bess Siding through Cloy, in which case it will throw the switch between Bess and Cloy and reenter the main track.

Lengths of DTC blocks vary but usually take about 10 minutes to traverse. The siding itself is not considered a DTC block per se, but in some situations sidings must not be occupied without authority from the train dispatcher. However, in DTC territory all sidings operate as non-controlled, restricted speed track and there is no safety consideration in a train occupying it without central authority to do so, however for traffic flow purposes the dispatcher need to track which sidings are occupied. If a train could not use the main track in Bess (due to maintenance work, a stuck switch, a disabled train, a fallen tree or boulder blocking the track, etc.), it would also be possible for the train to have authority from Anna through Bess Siding through Cloy. This is not done when the main track is available, as there is a speed disadvantage to having the train traverse the slower siding track (including having to stop to throw the switches to enter the siding).

While most DTC blocks are laid out back-to-back, the blocks actually do not touch at sidings. Instead, there is a “gray area” over the siding's switches, and the main track's and siding's DTC blocks begins at approximately the clearance point of the switch. The area over the switch must not be occupied unless authority is granted in blocks on both sides of the switch. For example, a train authorized only in Anna block must stop at the End Anna Block sign, which is a short distance from the switch points. A train authorized in Anna through Bess or in Anna through Bess Siding may proceed past the End Anna Block sign, over the switch, and then past the Begin Bess (or Bess Siding, as applicable) Block sign.

Blocks are typically named after a distinguishing feature within the block, such as a station, town, or river.

## Procedures

Normally, only one train or piece of equipment may occupy a given block at a time, and a train must release their authority in a block in order for the train dispatcher to authorize another train into that block. Barring human error, this ensures that no two trains are ever authorized on any given piece of track at the same time, thereby preventing collisions. (In certain circumstances, dispatchers will authorize more than one piece of equipment—as in the case of track maintenance vehicles—into the same block. When this occurs, the authority is marked as “joint,” and all movements must be made at restricted speed, which allows stopping within half the range of vision.) Another key feature of DTC is that only one authority is in effect for any given train at any given time. When a dispatcher issues a new authority to a train, the previous authority becomes invalid and must not be used.

In modern implementations, dispatchers rely on computerized systems to keep track of trains that have received authority. Typically, the computer will prevent the dispatcher from giving two trains authority over the same track. The computer system generally displays a highly simplified mock-up of the track, displaying the block limits and sidings. Track occupancy is displayed via bold or colored lines overlaying the track display, along with tags to identify the train (usually the number of the lead locomotive).

To receive or change a train's authority, the train dispatcher communicate by radio with the train crew and gives "mandatory directives" authorizing occupancy in a specified number of blocks. A member of the train crew, usually the conductor, copies the dispatcher's mandatory directive onto a prescribed form. The crew member will repeat the directive back to the dispatcher, who monitors the repeat for any errors. If the crew member correctly repeats the mandatory directive, the dispatcher finalizes the directive by saying, "That is correct," followed by the dispatcher's initials. A mandatory directive cannot take effect until the dispatcher has said, "That is correct," reducing the potential for human error to cause an accident.

Numbers are always repeated in an alternate format to ensure clarity and prevent misunderstandings. Numbers from zero to nine are spelled out, while numbers greater than ten are given by their individual numerals. For example, the number six would be given as, "Six, s-i-x," while the number 14 would be given as "fourteen, one-four." Engine numbers such as 2001 are usually read first as "two thousand one," while numbers such as 4321 are usually read as "forty-three twenty-one," although this is mostly a personal preference about what is easiest to say. Regardless, the repeat will give the individual numerals, as in, "Engine twenty-eight oh eight, two, eight, zero, eight".

A sample issue format would be as follows:

- Train dispatcher: "This is mandatory directive to engine forty-three twenty-one, four, three, two, one, with engineer Johnson. You are authorized to proceed westward in three, t-h-r-e-e blocks, Anna through Cloy, over".
- Train crew member: "This is a repeat of mandatory directive issued to the forty-three twenty-one, four, three, two, one, with engineer Johnson. We are authorized to proceed in a westward direction in three, t-h-r-e-e blocks, Anna though Cloy, over."

  Dispatcher: "That is correct, S.D.K., over.
- Crew: "That is correct, dispatcher S.D.K. Thanks, dispatcher, forty-three twenty-one, out."

  Dispatcher. "Thanks, train dispatcher out."

When trains must meet en route, one train must be authorized into the siding. For example, train 5432 may be authorized from Anna into Bess Siding:

- Train dispatcher: "This is mandatory directive to engine fifty-four thirty-two, five, four, three, two, with engineer Smith. You are authorized to proceed westward in two, t-w-o blocks, Anna through Bess Siding, over."

- Train crew member: "This is a repeat of mandatory directive issued to the fifty-four thirty-two, five, four, three, two, with engineer Smith. We are authorized to proceed westward in two, t-w-o blocks, Anna through Bess Siding, over."
- Dispatcher: "That is correct, S.D.K., over.
- Crew: "That is correct, dispatcher S.D.K. Thanks, dispatcher, fifty-four thirty-two, out."
- Dispatcher. "Thanks, train dispatcher out."

An opposing train will be authorized in two blocks, Cloy through Bess, using a similar format:

- Train dispatcher: "This is mandatory directive to engine forty-three twenty-one, four, three, two, one, with engineer Johnson. You are authorized to proceed eastward in two, t-w-o blocks, Cloy through Bess, over."
- Train crew member: "This is a repeat of mandatory directive issued to the forty-three twenty-one, four, three, two, one, with engineer Johnson. We are authorized to proceed eastward in two, t-w-o blocks, Cloy through Bess, over."
- Dispatcher: "That is correct, S.D.K., over.
- Crew: "That is correct, dispatcher S.D.K. Thanks, dispatcher, forty-three twenty-one, out."
- Dispatcher: "Thanks, train dispatcher out".

Once train 5432 has fully entered the siding and restored the switch for movement along the main track, they will call the dispatcher and release Anna block, using the following format:

- Train crew member: "The fifty-four thirty-two, five, four, three, two, would like to release our authority in one block, Anna, over".
- Train dispatcher: "Engine fifty-four thirty-two, five, four, three, two, is releasing authority in one block, Anna, over."
- Crew: "That is correct, dispatcher."
- Dispatcher: "All right, thanks. Dispatcher out."
- Crew: "Fifty-four thirty-two out".

After the 5432 has released its authority in Anna, the dispatcher may issue a new authority to the 4321 authorizing them through Anna block. Since only one authority is valid at any given time, if the 4321 is still in Cloy block, the dispatcher will give the new authority containing Cloy through Anna. Once the 4321 is fully within the Bess block and has cleared Cloy block, the conductor or other crew member will call the dispatcher and release Cloy. The dispatcher will then issue a new authority to the 5432, giving them authority from Bess Siding block through Cloy block. This generates additional radio traffic but ensures that there is no confusion as to which train is allowed to enter a given block at any given time. Some railroads have instituted a practice which allows a train to enter a block contingent upon the passing of the opposing train, although the National Transportation Safety Board has recommended against this practice. A DTC authority given under this practice will include a phrase such as "This line is not in effect until after the arrival of train 7940".

In the case of two trains traveling the same direction, the dispatcher will only authorize the following train in blocks that the leading train has released. In the case of closely spaced trains, the dispatcher may opt to reduce his or her workload and implement a "radio blocking behind" authority, which lets the two trains talk to each other directly to discuss which blocks the leading train has cleared. Caution must be exercised by the leading train not to release authority on track that the train still occupies, as this could result in a rear-end collision if the leading train slows down or stops or if the train becomes separated and a portion is left standing on the main track.

Entering a block in which the train is not authorized or releasing a block which the train still occupies is considered a main-line violation by the Federal Railroad Administration and carries stiff penalties for all crew members, which may include time off from work and the suspension or revocation of the engineer's license.

### Use

The Southern Pacific Railroad, now the Union Pacific, implemented DTC on several of its lines formerly controlled by train orders (most of the rest of UP's lines use Centralized Traffic Control). The BNSF Railway does not use DTC; it uses mostly CTC with track warrants in its non-signalled territory. The Alaska Railroad changed its Track Warrant Control territory to DTC and is also implementing CTC in high-traffic areas. Queensland Rail in Australia uses the DTC system in some of the lower traffic areas in the western area of the state. CSX uses DTC on their former Georgia railroad mainline from Scottdale, just east of Atlanta, to Augusta.

## Train Order Operation

Train order operation, or more accurately timetable and train order operation, is a largely obsolete system by which the railroads of North America conveyed operating instructions before the days of centralized traffic control, direct traffic control, and the use of track warrants conveyed by radio. The system used a set of rules when direct communication between train dispatchers and trains was limited or non-existent. Trains would follow a predetermined operating plan, known as the timetable, unless superseded by train orders conveyed to the train from the dispatcher, through local intermediaries. Train order operation was a system that required minimum human overhead in an era before widespread use of technology-based automation. It was the most practical way for railroads with limited capital resources, or lines with limited traffic, to operate. To this day, a large number of short lines, heritage railways, and railroad museums continue to use Train Order operation.

On major railroads, train order operation has been almost completely replaced by more modern operating methods. The Long Island Rail Road in New York is the last major railroad in North America to use a "traditional" Train Order operating practice on parts of its Greenport and Montauk Branches, as well as Train Order forms for non-standard operation on the remainder of its system. While the last traditional long hand train order form was issued on September 3, 2012, timetable and train order operating practices remain in effect. The second to last train order holdout, the Chicago South Shore and South Bend Railroad, had retired its system in 2011.

The process of modernization in the 19th century involved a transition from a spatially oriented world to a time oriented world. Exact time was essential, and everyone had to know what the time was, resulting in clocks towers for railway stations, clocks in public places, pocket watches for railway workers and for travelers. Trains left on time (they never left early). By contrast, in the premodern era, passenger ships left when the captain had enough passengers. In the premodern era, local time was set at noon, when the sun was at its highest. Every place east to west had a different time and that changed with the introduction of standard time zones. Printed time tables were a convenience for the travelers, but more elaborate time tables, called train orders, were even more essential for the train crews, the maintenance workers, the station personnel, and for the repair and maintenance crews, who knew when to expect a train would come along. Most trackage was single track, with sidings and signals to allow lower priority trains to be sidetracked. Schedules told everyone what to do, where to be, and exactly when. If bad weather disrupted the system, telegraphers relayed immediate corrections and updates throughout the system. Just as railways as business organizations created the standards and models for modern big business, so too the railway timetable was adapted to myriad uses, such as schedules for buses ferries, and airplanes, for radio and television programs, for school schedules, for factory time clocks. The modern world was ruled by the clock and the timetable.

## Train-order Station

A train-order station is a control point at which trains can be stopped and controlled through the use of train orders. A station has a distinct name, and may have any of the following:

- A siding or other track by which trains can pass each other.
- A communications means for an operator to receive train orders.
- A signal to indicate to trains whether there are train orders to be picked up.

A train order station need not be at a passenger or freight station, nor does such a station have to handle train orders. In isolated areas, train order stations may be required where there are no towns, to facilitate smooth operation. In denser areas, passenger stations may be spaced more closely than train order stations.

A station may staffed by an operator who receives train orders and gives them to trains as they pass. Operators also record the passage of trains by their station. Upon receipt of an order, the operator makes copies and sets the signal to indicate to approaching trains that orders are to be picked up. Some train-order signals had three positions:

- Proceed (green light or vertical blade)

  No orders; train may proceed
- Receive orders (yellow light or diagonal blade)

  Pick up orders without stopping
- Stop (red light or horizontal blade)

  Stop to receive orders or to wait for another train to pass

# Application of Railway Signals

The application of railway signals on a rail layout is determined by various factors, principally the location of points of potential conflict, as well as the speed and frequency of trains and the movements they require to make.

## Non-provision of Signals

Before discussing the application of signals, it is useful to highlight some situations where signals are *not* required:

- Sidings generally need not be signalled, as all movements within them are made cautiously at low speed.
- Tram lines frequently employ "running on sight" without any signals (similar to road traffic).
- Where movement authorities are passed to drivers exclusively by means other than fixed signals (e.g. by written or verbal authority), token, or cab signaling.

## Purpose of Signals

Signals exist primarily to pass instructions and information to drivers of passing trains. The driver interprets the signal's indication and acts accordingly. The most important indication is 'danger', which means 'stop'. Not every signal is equipped with a 'danger' aspect.

## Protection of Conflict Points

Signals are provided where required to protect items of infrastructure where conflict may arise, including:

- Points (switches).
- Moveable bridges.
- Certain types of road level crossings.
- Level junctions and gauntlet track.

A signal cannot display a 'proceed' aspect unless the infrastructure item(s) that it protects are in the correct position for the passage of a train or, in the case of flat junctions and gauntlet track, no other signal is cleared for a conflicting move. This is enforced by interlocking.

## Protection of Following Trains

Running lines (as distinct from sidings) are divided into sections. Under normal circumstances, only one train may occupy any section at a time. A signal is provided at the start of every section, which may only display a 'proceed' aspect when the section ahead is completely empty, at least as far as the next signal.

The length of the sections, and hence the distance between signals, determines the railway's capacity. A railway with short sections can accommodate more traffic than one with long sections.

## Signaling a Layout

The provision of signals is dependent on the intended use of the layout and the movements that are expected to take place. It is not necessary to provide a signal for every conceivable movement that the layout physically permits. If, during perturbed working, a movement has to be made in a direction for which no signal is provided, this can be done under special instruction. The same applies during signal failure.

## Signal Categories

### Main Signal

A main signal controls a train movement along a running line main line. A 'proceed' aspect on a main signal indicates that the line is clear at least as far as the next signal, giving the driver confidence to run at speed. Trains running long distances, especially passenger trains, will usually travel throughout under the authority of a succession of main signals.

### Shunting Signal

A shunting signal controls low speed movements where provision of a main signal is not appropriate, such as moves into sidings. Clearance of a shunting signal does not necessarily imply that the line ahead is clear of vehicles. Shunting signals are often mounted at ground level and are smaller than main signals, reflecting their status.

### Stop Signal

A stop signal is one that is equipped to show a 'danger' aspect, which commands an approaching train to stop. Its function is to prevent conflict with other trains and to indicate that moveable infrastructure features are in the correct position. Depending on the manner in which they are used, and the nomenclature favoured by the railway administration concerned, stop signals may be further categorised as 'home signals' or 'starting signals', for example. Some stop signals are in the form of a fixed signal, generally a white board with a red solid circle. There is usually a panel underneath with instructions to the driver as to what circumstances he may pass the signal. Examples of such signals were used on lines signaled by the radio electronic token block system.

### Distant Signal

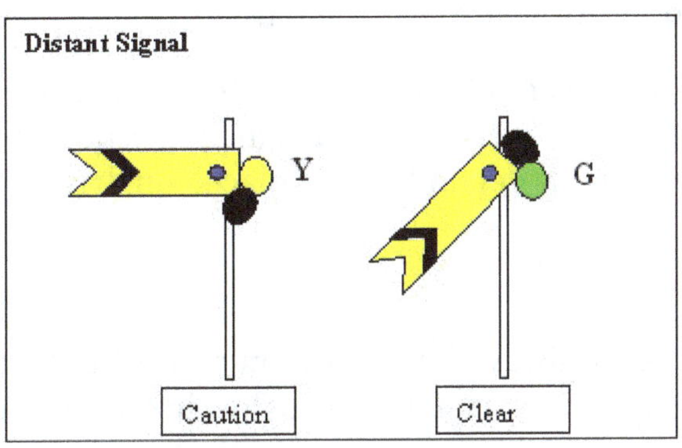


Mechanical distant signal.

A distant signal is one that cannot display a 'danger' aspect. It is however able to display a 'caution' aspect, which gives the driver advance warning that the stop signal ahead may be displaying 'danger'. The distant signal is installed roughly at full braking distance on approach to the stop signal to which it applies, taking into account the gradient, the permitted speed, and the braking performance of trains. On sighting a 'caution' aspect, the driver must prepare to stop at the signal ahead. If the distant signal shows a 'clear' aspect, the train may maintain full speed. A single signal may be equipped to function both as a stop signal and a distant signal. Some distant signals are in the form of a 'fixed distant'. That is: they only ever display a 'caution' aspect and never a 'clear' one. Such signals are usually in the form of a standard distant arm fixed in the horizontal position to the signal post. An alternate form is a picture of a horizontal distant arm painted onto a white board.

### Permissive Signal

The 'danger' aspect of a permissive stop signal means "stop and proceed". The driver is permitted to pass the signal at 'danger' under his/her own authority in accordance with the rules, after first coming to a stand.

### Subsidiary Signal

A subsidiary signal permits movements onto a portion of track that is already occupied. These are commonly used at terminal stations to permit two or more trains to enter a single platform.

### Junction Signaling

The driver of a train approaching a diverging junction must be made aware of which direction the train will take, so that its speed can be regulated accordingly. A diverging route might have a significantly lower permissible speed than the main route, and if the route taken was not the one expected, it could result in derailment.

There are two different philosophies on junction signaling. Signaling of British origin generally conforms to the principle of *route signaling*. Most railway systems around the world, however, use what is known as *speed signaling*.

Under route signaling, a driver is informed of which route the train will take beyond each signal (unless only one route is possible). This is achieved by a 'route indicator' attached to the signal. The driver uses his route knowledge, reinforced by speed restriction signs fixed at the lineside, to drive the train at the correct speed for the route to be taken.

Under speed signaling, the driver is not told which route the train will take, but the signal aspect informs him at what speed he may proceed. Speed signaling requires a far greater range of signal aspects than route signaling, but less dependence is placed on drivers' route knowledge.

### Bidirectional Signaling

Many double or multiple track railways have a designated direction of travel assigned to each track. Signals will only be installed to permit traffic to flow in one direction on each line (i.e. uni-directional signaling).

Bidirectional signaling in evidence at Poyntzpass with a passing Enterprise on NIR.

Bidirectional signaling is the provision of signaling that allows one or more tracks on a multiple track railway to be operated in either direction, whether for regular or emergency use. Bidirectional signaling intended for regular use will generally allow traffic to flow at similarly high frequency in one or other direction. If intended for emergency use, running in the opposite direction might only be possible at reduced frequency. Typically, 'reduced capacity' bidirectional signaling only provides a signalled route onto and off the other line for 'wrong' direction running, without any intermediate stop signals that would improve capacity.

Bidirectional signaling is more expensive to implement as it requires more equipment than uni-directional operation, so it is not always provided. In the absence of bidirectional signaling, 'wrong direction' movements may still be made on a uni-directional line at times of disruption, through a procedure known as single line working.

Bidirectional signaling is easier to implement when under the control of one signal box. Before power-operated signaling became widespread, track layouts tended to be designed to avoid bidirectional arrangements as much as possible.

Single track railways must necessarily have signals for both directions. This is not regarded as bidirectional signaling. However, some crossing-loops are fitted with bi-directional signaling to allow trains (usually when no crossing is to be made) either to pass at speed on a straighter route or to serve a more convenient platform for passengers.

## Signals for Special Purposes

### Train Order Signal

Used in conjunction with the American train order system, a train order signal advises the engineer of the need to pick up a train order at a station.

### Hump Shunting Signal

In a hump yard, special signals may control the speed of trains propelling vehicles towards the hump.

### Loading and Unloading Signals

Signals may be installed to control the movement of freight trains through a facility for loading or unloading minerals, for example coal. Unlike ordinary signals, there may be a series of identical signals installed along the track so that at least one is always visible to the train driver at any time. All signals in the same group display the same indication simultaneously. A 'stop' indication means "stop *immediately*", even if the train is not at the signal. The signals can also instruct a train to reverse.

## References


- Railroad-signals, abcs-of-railroading, railroads: trn.trains.com, Retrieved 10 February 2019
- United States Federal Railroad Administration (2009-02-20). "Positive Train Control Overview". Archived from the original on 2010-02-19. Retrieved 2010-10-05
- Rail-Traffic-Management-and-Signaling: navipedia, gssc.esa.int, Retrieved 11 March, 2019
- Daniels, Rudolph (2000). Trains Across the Continent: North American Railroad History. Indiana University Press. P. 223. ISBN 0-253-21411-4
- Briginshaw, David (January 8, 2014). "JR East selects Thales to design first Japanese CBTC". hollandco.com. Holland. Retrieved January 9, 2014

# 5
# Train Protection

Train protection refers to the system which ensures the safe movement of trains. The train protection systems are broadly classified as two channel safety systems and automatic train protection systems. All these diverse types of train protection systems have been carefully analyzed in this chapter.

Railway signaling is the basic safety system controlling the movements of trains. It is the safety critical part of the train control function of the railway. Once instructed by a signaller or an automatic system, it is responsible for setting up non-conflicting and safe routes for trains, for defining (safe) limits of movement and for transmitting instructions or commands to train drivers.

There are two principal parts to a train protection system, train detection (knowing where the train is) and movement authority (telling the train how far it can go). The train protection system uses these two parts to safeguard a train's operation.

Traditionally, signaling systems in Britain and in many other countries have relied on the train drivers reacting to indications displayed to them by line-side semaphore or colour light signals and controlling the train's speed in line with the instructions. During the 150 years of the use of railway signaling, drivers' failures to respond to commands transmitted by signal aspects of any type can and have led to a number of accidents, some causing very large numbers of fatalities. In response to the continuing need to reduce risks created by train drivers failing to respond to signal instructions, various forms of driver warning devices and signal command enforcement systems have been developed. These have become known as Train Protection Systems. Those systems that continuously monitor actual train speed and enforce adherence to a commanded speed pattern are referred to as Automatic Train Protection (ATP) systems.

## Types of Train Protection Systems

All types of train protection systems are based on the desire to reduce or eliminate the possibility of driver error resulting in a train movement related accident by failing to obey a visually displayed line-side or in-cab signal instruction. The development of train protection on main line railways began with the introduction of warning systems and subsequently progressed to enforcement of the instructions issued by these systems.

Originally, the warning systems alerted the driver that he or she was approaching an adverse or restrictive line-side signal aspect and required him or her to acknowledge the warning. Otherwise the systems would initiate a brake application after a short delay. Later developments by national railway administrations included various levels of speed limitation and enforcement. Also, some systems were expanded to cater for speed limits for permanent or temporary speed restrictions. Technologies adopted for such warning and train stop systems include combinations of permanent magnets and electromagnets, inductive polarity-changing responders, coded beacons and simply coded track circuits.

More recently, fully Automatic Train Protection (ATP) systems have been developed to enforce speed limits and movement authorities at the full range of restrictive signals, with and without line-side signals and including permanent and temporary line speed limits. Driving is still manual but speed limits are always enforced. Degraded modes though invariably include low speed driving on sight.

## Two-channel Safety Systems

Many older train protection systems are designed to rely on the statistical nature of driver and equipment failure. By designing the systems appropriately, it may safely be assumed that driver error and equipment failure will not occur simultaneously. A key characteristic of such systems is that the driver does not receive an indication of whether the train protection system is operating or not and is therefore encouraged to drive taking full responsibility for the movement of the train. The technical subsystem will only intervene if the driver attempts to pass a signal or to drive too fast. TPWS, train stop and Indusi are typical examples of this type of arrangement.

## Automatic Train Protection Systems

There are broadly two implementations of ATP systems – intermittent and continuous. Intermittent systems use electronic beacons (inductive or radio frequency) or short electrical loops positioned within the four-foot. These types of short-range devices are often referred to as "balises" (from the French word for 'marker'). The continuous systems use a permanently active data transmission and monitoring system, either through electrical inductive coupling by means of track loops or coded track circuits or by means of radio transmission of limit of movement authorities.

Fully operational ATP systems were first introduced on metros in the late 1960s and are now common on such systems all over the world. Most metro applications use continuous systems in conjunction with automatic train operation. ATP was also introduced to the Japanese Shinkansen high speed route in 1964 and has since been introduced in various forms on a number of main line railways, often in conjunction with high speed train operation.

## Principles

The basic defining principle of ATP is that train speed is monitored against the current permitted speed limit. The speed may be limited by line profile or signal indication, that is, the need to protect routes of other trains and track related constraints. If the allowable speed is exceeded, a brake application is invoked until the speed is brought within the required limit or the train is stopped.

Most ATP systems are based on conventional block signaling although these can be very short. Each block is described by a fixed dataset related to its location, length, gradient(s) and maximum (civil) speed limit(s). Each block will also have a variable data set derived from the signal aspects ahead and their effect on the resulting speed limit(s) for that block and the next block(s).

## Enforcement

The speed limit on the approach to a restrictive signal forms a gradually reducing curve that follows the braking profile required to reach the target speed at the signal. If the signal shows a stop aspect, the target speed will be zero. The on-board monitoring equipment will continuously compare the train speed with the curve required to achieve the target speed and will initiate a warning – usually both audio and visual. If action is not taken, the system will cause a brake application.

Photo of a track mounted balise and the train mounted data reader. The balise may be passive or active. Passive balises provide fixed data such as train location updates or permanent speed limits. Active balises provide variable data concerning signal aspects and the associated limit of movement authority.

In some implementations, a braking curve infringement calls for a full service brake initiation, in others for emergency brake. There are also differences in the brake release function. Some systems allow the driver to release the brake once the train speed has returned within the prescribed curve. In others, the brake command is irrevocable and the train must be brought to a stand before the driver can release the brake. There are also railway undertaking specific rules about the consequences when the ATP system has intervened.

## Train Data

On the train, data comprising train weight, length, braking capability and maximum technically permitted speed are necessary to ensure compliance with speed limits set by the ATP system. Usually, the train consist data must be input by the driver before the trains starts its journey. Ways of validating this may be considered necessary.

## Monitoring

In most cases, the performance of the equipment is monitored and recorded for further analysis in case of infringements or failures of the system. These systems are variously known as On Train Monitoring Systems (OTMS), On Train Data Recorders (OTDRs) or On Train Monitoring Recorders (OTMRs). They are the equivalent of the aircraft industry's 'black box'.

## Drawbacks of Intermittent Systems

Continuous ATP systems allow constant data updates to be transmitted to trains so that the train driver can respond to changes in signal aspects as soon as they occur. Intermittent systems can only transmit changes in signal aspects when the train passes over a beacon or loop. This can restrict line capacity if a driver is unable to respond to a signal clearance, even though he or she can see the change of aspect, until the train's on-board ATP computer has received a message from the balise located at the relevant signal.

In order to overcome this problem, infill loops or balises are provided at some signals (e.g., in 65% of the locations on the UK GWML system) to provide drivers with an update of a signal aspect and to allow brake release if a less restrictive aspect is shown.

## Automatic Warning System (AWS)

Following a Signal Passed at Danger (SPAD) accident in poor visibility at Harrow and Wealdstone in 1952 when 112 persons were killed, British Railways decided to deploy their Automatic Warning System (AWS) over the whole network to provide train drivers with an in-cab warning of the indication of the next signal (Figure). This was a non-contact version of a system originally used on the Great Western Railway. Following a long development and approval programme, widespread installation started in 1956. This system is still in use today.

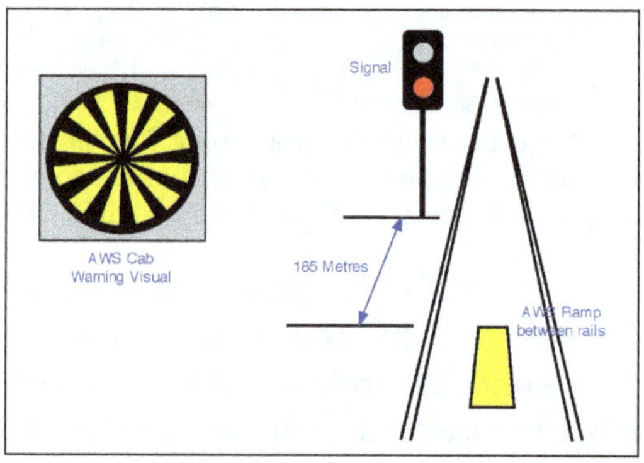


Schematic showing arrangement of AWS Ramp on the approach to a signal. When a train passes over the ramp, an indication of the signal aspect is provided in the driver's cab. In the case of a caution or stop aspect, the system will enforce an application of the brakes unles the indication is acknowledged by the driver.

The system took the form of a track mounted, non-contact inductor, which became known as the AWS " amp", placed about 185 metres (200 yards) on the approach side of the signal.

The AWS ramp is placed between the rails so that a detector on the train will pass over it and receive a signal. The ramp will thus warn the driver of the status of the signal. The French railways use a similar system called "the Crocodile", the Germans, the "Indusi".

The AWS ramp contains a pair of magnets, the first permanent, the second an electro-magnet linked to the signal to provide an indication of the aspect. The ramp is placed between the rails so that a detector on the train can receive the indication data. The more observant passenger on a station platform can often see the ramps between the rails. They are usually a dirty yellow.

In operation, the train first passes over the permanent magnet and the on-board receiver sets up a trigger for a brake application. Next it passes over the electro-magnet. If the signal is green the electro-magnet is energised, the brake trigger is disarmed, a chime or bell rings in the driver's cab and a black indicator disc is displayed. The driver takes no action. If the signal is yellow or red, as shown in figure, the electro-magnet is de-energised, so a siren sounds in the cab and the disc becomes black and yellow. The driver must "cancel" the warning, otherwise the automatic application of the train brakes is triggered.

Photo showing position of AWS Ramp in the track on the approach to a signal.

The use of AWS was extended to protect certain permanent speed restrictions after an accident at Morpeth in 1969, which was caused by a driver failing to reduce the train speed for a speed restriction. AWS was installed at speed restriction locations where the train approach speed is over 60 mph or the reduction in speed required is more than one third of the approach speed. This is mandated in a railway group standard.

After an accident at Nuneaton in 1975, the use of AWS was extended further to include temporary speed restrictions. It is now also used for emergency speed restrictions.

Although AWS has been partially effective in reducing train movement accidents, it has not eliminated SPAD or over-speed errors entirely. By design, it is only intended to provide an alert and a reminder of a restrictive signal aspect or speed restriction. As long as the alert is acknowledged, the driver may continue to drive the train at any speed. A number of accidents occurred where drivers forgot the restrictive aspect warning – despite the visual reminder in the shape of the so-called 'sunflower' device.

AWS may be considered to a limited extent as a fail-to-safety system since the main trigger element is a permanent magnet. Failure of the electro magnet results in a warning indication to the driver. However, removal of the permanent magnet from the track is not detected.

## Driver's Reminder Appliance (DRA)

The Driver's Reminder Appliance (DRA) was introduced from 1998 to assist with the prevention of SPADs, particularly at station starting signals. It is not really a train protection device in the narrow sense of the terminology. The DRA device consists of a push button in the driver's cab. The driver is required to operate this whenever the train is detained at a red signal. The button disables traction and prevents the driver from restarting the train until he has reset the button. The primary purpose is to prevent trains starting against a red signal when inadvertently given the "right away" by a conductor or station staff. The effectiveness of this system is a matter of debate because its operation may become 'automated' as part of the train starting sequence.

## Train Protection and Warning System (TPWS)

To overcome the limitations of AWS, an enforcement system was developed for the British railway system, known as TPWS (Train Protection and Warning System). This is designed to enforce observance of restricted speed requirements and signal stops by imposing a full brake application when over-speed is detected or when a train is being driven past a stop signal. TPWS was first tried on a section of the Thameslink route in 1996 and was then installed across most of the UK network between March 2000 and December 2003.

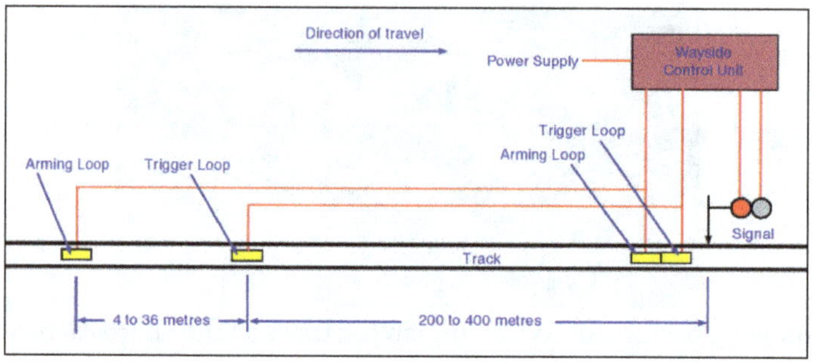


Schematic of TPWS setup on the approach to a stop signal. The Arming Loop switches on a timer and the Trigger Loop assesses the time elapsed to determine the speed of the train. If the time is too short, showing the speed is too high, the trigger will activate the train brakes.

The idea behind TPWS is that, if a train approaches a stop signal showing a danger aspect at too high a speed to enable it to stop at the signal, it will be forced to stop, regardless of any action (or inaction) by the driver. The equipment is arranged as shown in figure.

For each signal equipped with TPWS, two pairs of electronic loops are placed between the rails, one pair at the signal itself, the other pair some 200 to 450 metres on the approach side of the signal. Each pair consists of, first an arming loop and secondly, a trigger loop. The loops are activated if the signal is showing a stop aspect.

The pair of approach loops first met by the train at 400 to 200 metres before the signal, are set between 4 and 36 metres apart. When the train passes over the arming loop, an on-board timer is switched on to detect the elapsed time while the train passes the distance between the arming loop and the trigger loop. This time period provides a speed test. If the test indicates the train is travelling too fast, a full brake application will be initiated. In case the train passes the speed test successfully at the first pair of loops but then fails to stop at the signal, the second set of loops at the signal will cause a brake application. In this case, both loops are together so that, if a train passes over them, the time elapsed will be so short that the brake application will be initiated at any speed.

## What TPWS Does

TPWS has certain features which allow it to provide an additional level of safety over the existing AWS system but it has certain limitations and does not provide the absolute safety of a full Automatic Train Protection (ATP) system. What TPWS does is reduce the speed at which a train approaches a stop signal if the driver fails to get the speed of the train under control to allow him to stop at the signal. If the approach speed is too fast, TPWS will apply a full brake but the train may still overrun the signal. Fortunately, since the train is already braking and there is usually a "cushion" of 200 yards (183 metres) between the signal and the block it is protecting, there will be a much reduced risk of damage (human and propertywise) if the train hits anything. With a possible total distance of 2000 feet (about 600 m) between the brake initiation and the block entrance, trains "hitting" the first loops at up to 120 km/h (75mph) could be stopped safely.

Pair of TPWS grids on the approach to a stop signal at a station platform.
The AWS ramp for the same signal is also visible. The distance between the TPWS grids provides the time difference used to check the train speed to ensure it is low enough to stop at the signal.

TPWS is also provided at many Permanent Speed Restrictions (PSRs) to ensure that a train does not pass through a restricted section of line (say one with a sharp curve) at too high a speed. However, there have been a number of issues related to the use of TPWS in these cases. Drivers have complainted that, although they were approaching the PSR at a speed which would allow the train to run at the correct speed within the restriction, they still got stopped by the TPWS "speed trap". This has led to some vigorus discussions between Network Rail, the train operating companies and the HSE. An add-on to TPWS, called TPWS+ is provided at certain signals where train speeds are above 100 mph or 160km/h.

TPWS does not replace the existing AWS system. AWS is retained, so the driver will still get the warnings advising him of adverse signals. The TPWS equipment is designed to interface with the existing on-board wiring of trains so that it can be fitted quickly.In an attempt to reduce the SPAD risks at certain high-speed locations, an add-on to TPWS, called TPWS+ is provided at certain signals where train speeds are above 100 mph. An additional loop pair is set about 770m in rear of the signal in order to provide the braking distance for a train "tripped" at 100 mph. Therefore, in these locations the signal approach has two speed traps. There are about 500 sites chosen on a risk assessment basis.

A further variation of TPWS, designed to be compliant with European requirements and known as TPWS-E, was tried on a section of the British Great Western Main Line (GWML) but it was not proceeded with further in order to allow rapid deployment of the already approved TPWS equipment.

TPWS is also provided at many (about 3000) Permanent Speed Restrictions (PSRs) to ensure that a train does not pass through the section of line at too high a speed. TPWS has also been provided at terminal platforms to ensure the train speed is reduced to 15 mph on the approach to the stops. This has had the effect of reducing capacity at some terminals because of the time taken for trains to clear the routes over the throat into the terminus. This type implementation may also encourage drivers to re-motor when travelling along a platform.

In spite of the limitations of TPWS, it is suggested in published data that 60% of potential accidents due to SPADs can be prevented by the installation of TPWS at critical locations. This has been achieved, it is said, at 10% of the installation costs of a full ATP system. However, this financial target was not achieved because of a decision to monitor the status of the TPWS beacons. With the lack of any in-built failure warning capability for TPWS, it was decided to link installations to the signal in rear, so that, in the event of a TPWS failure, the signal would display a red aspect. This addition to the original specification significantly increased costs.

TPWS does not replace the existing AWS system. AWS is retained, so that drivers still receive the warnings advising of adverse signal indications. The TPWS equipment was designed to interface with the existing on-board AWS equipment on trains or to replace it so that it could be fitted quickly.

## Radio Electronic Token Block (RETB)

In some rural areas of the UK, where long sections of single line require token block operation, a system for centralised control, using modern computer technology, was adopted. It is known as Radio Electronic Token Block (RETB).

Each train operating over the single line is equipped with a special speech and data radio transmitter/receiver with a unique identity. At the start of the single line, the driver stops and calls the control centre for authority to enter the section. If the line is clear, the signalman in the control centre transmits a coded "electronic token" data message which is received by the train and then shows the authority for that section on a cab display. The driver will then call for confirmation that he can enter the section. Once in the single line section, he will advise the control centre that he has cleared the loop track. A clearance marker board is specially provided to help him. When he has reached the end of the single line section, the driver calls the control centre again and offers to give up the token. After a "handshake" procedure by the control centre, he sends the token back by radio data transmission to release the section.

The signaller is provided with a computer system that allocates the coded tokens to each section and prevents more than one token being issued for an occupied section. It also receives the tokens sent back by each train as it reaches the end of the single line section.

At the exits of the single line sections, the points are permanently set in the direction of normal running and are "trailable" for trains entering the section, i.e. they allow a train to pass through at reduced speed using the wheel flanges to move the point (switch) blades aside reset to the normal position.

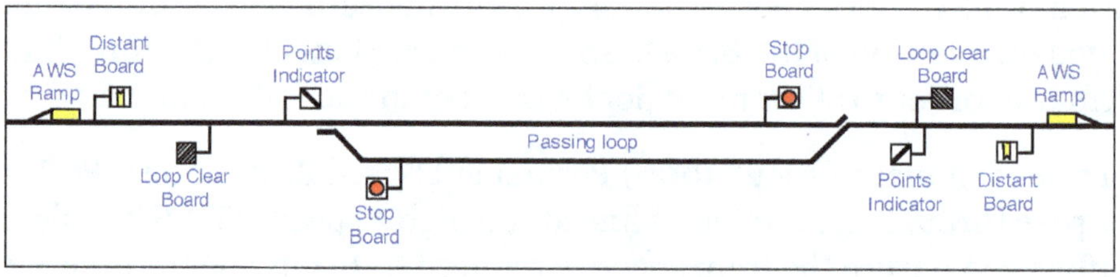


Diagram of route with Radio Electronic Token Block system.

A "Distant Board" complete with AWS ramp, warns the driver that he must slow down for the movement over the points leaving the single line. The Points Indicator shows the position of the points. A "Stop Board" at the end of the passing loop warns the driver to stop and ask for permission to enter the next single line section. Stop board locations are provided with TPWS loops that are linked by radio to the signal controller. A "Loop Clear" board indicates to the driver when the rear of the train is clear of the points.

An excellent description of the operation of RETB is available at "The Modernisation of the Cambrian Lines" by Alan Jones on the Signal Box Site. This system has been superceded by the test installation of ERTMS on the route. The East Suffolk Line's system was converted to conventional track circuit block with axle counters when the service frequency was increased to a level where RETB could not have handled it. It was phased out in October 2012. RETB remains in use on several of the more remote routes in Scotland.

# Automatic Train Control

A railway train is an ideal system for automation. It uses a fixed guidance system, its acceleration and braking can be predicted, its position detected, its direction confirmed and its timing regulated. All this makes automation of train control a relatively simple task. However, there are limitations:

- Train formations that can vary need to be individually registered into the system;
- Variations to railhead conditions need to be factored into the system;
- Not all existing railways are ideal for automation and may need significant upgrades.

Nevertheless, automation has considerable benefits for safety and performance and can offer better throughput of trains of up to 8% just by the elimination of manual driving variability.

The automation of train movements was gradually developed from the need to enforce signal commands so that drivers could not allow trains to pass beyond their limit of movement authority (LMA). The automation of train control was developed from there, using advances in traction control, which gave automatic acceleration and in braking, which could incorporate electrical control and automatic load compensation.

Originally, in Britain, the letters ATC referred to "Automatic Train Control", which was the title given to the warning system tried on some UK lines before the general introduction of the AWS (Automatic Warning System) in the 1960s. In the US it also refers to Automatic Train Control but it refers to a more modern concept where the system includes ATP (Automatic Train Protection), ATO (Automatic Train Operation) and ATS (Automatic Train Supervision). ATC has been adopted around the world to describe the architecture of the automatically operated railway. It is usually applied to metros.

As a definition, ATC refers to the whole system which includes all the other automatic functions and, for some of these functions at least, also includes a degree of manual intervention. ATC therefore, is the package which includes ATP, ATO and ATS.

## The ATC Package

The first part of the package of an ATC system is Automatic Train Protection (ATP), where the train is given a Limit of Movement Authority (LMA). This is based on the train's current speed, its braking capability and the distance it can go before it must stop. On a manually driven train, the driver manages this through a combination of route and stock knowledge and the visual information received from wayside signals. On the automated train, the data for the LMA is transmitted from the track to the train where the on-board computer registers the current speed and calculates the target speed that the train must reach and by when. This is electronically plotted in the form of a braking curve. If the train is allowed to exceed the profile of the curve, the brakes will automatically apply to bring the train to a stand (or at least to within the permitted speed).

The second part of an ATC system is Automatic Train Operation (ATO). This is the driving part of the operation. Looking at a manually driven train, we will see that the driver initiates the starting of the train, allows its acceleration to the permitted speed, slows it where necessary for speed restrictions and stops at designated stations in the correct location. The ATO system will carry out these parts of the operation with the exception that the driver normally initiates the train start. There are a number of different systems used to perform the ATO functions but they all involve data communication between the train and the train and most require some form of on-board route map.

There are a number of ways to assemble the parts of an ATC package but a common format appears as in figure.

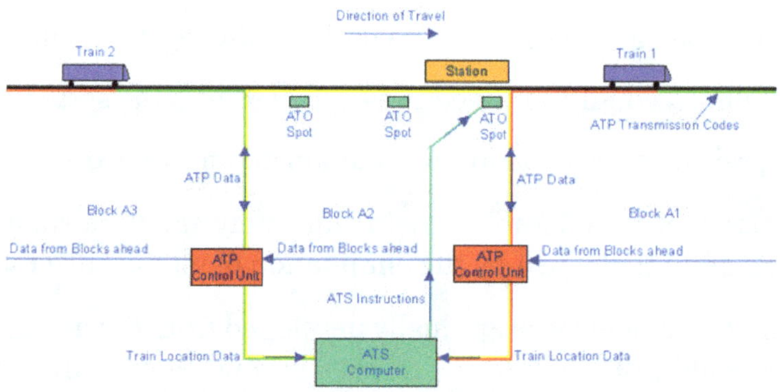


Overview of ATS as applied to a fixed block system

A schematic showing the basic architecture of a fixed block automatic train control (ATC) system with its three main components - ATP (Automatic Train Protection), ATO (Automatic Train Operation) and ATS (Automatic Train Supervision).

The data received by the ATP control unit is usually limited to indicating that a train is in the block or the speed limit currently imposed in the block. This data is sent to the ATS computer where it is compared with the timetable to determine if the train is running according to schedule or is late or early. To adjust the train's timing, the ATS can send commands to the ATO spots located along the track.

The ATO spots, which can be short transmission loops or small boxes called beacons or "balises", give the train its station stop commands. The spots usually contain fixed data but some, usually the last one in a station stop sequence, transmit data about the time the train should stop (the dwell time) at the station and may tell it how fast to go to the next station (ATP permitting).

Some systems leave the ATO spots alone - i.e their data is always fixed - but use the ATP system to prevent the train from starting or restrict its speed. The ATS computer tells the ATP control unit to transmit a restricted speed or zero speed to the track.

Both ATP and ATO commands are picked up by receivers on the train and translated into motoring, braking or coasting commands. Where a train can be manually driven, the ATP will still ensure the safety requirement but the ATO is overridden, the driver stopping the train in the stations by use of the cab controls.

There are lots of variations of ATC around the world but all contain the basic principle that ATP provides safety and is the basis upon which the train is allowed to run. ATO provides controls to replace the driver, while ATS checks the running times and adjusts train running accordingly.

## The Overlap

If a line is equipped with a simple ATP system which automatically stops a train if it passes a red signal, it will not prevent a collision with a train in front if this train is standing immediately beyond the signal.

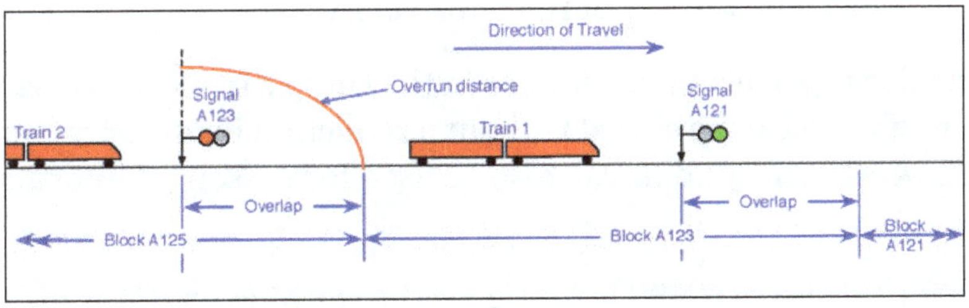


Diagram showing a signal provided with an overlap. The overlap on British main lines was originally based on statistics showing how far a train would overrun in misjudging the stop at a signal. On the London Underground they were calculated for each location. Many railways use a complete block as a safety margin.

There must be room for the train to brake to a stop. This is known as a "safe braking distance" and space is provided beyond each signal to accommodate it. In reality, the signal is placed in rear of the entrance to the block and the distance between it and the block is called the "overlap". Signal overlaps are calculated to allow for the safe braking distance of the trains using this route. Of course, lengths vary according to the site; gradient, maximum train speed and train brake capacity are all used in the calculation.

Overlaps are often provided on main line railways too. In the UK, it is the practice to provide a 200 yard (185 m) overlap beyond each main line signal in a colour light installation. Back in 1972 when it was decided upon, it was, after a review of many instances where trains had overrun stop signals, considered the maximum normally required. It was a rather crude risk analysis but it was the best they could afford.

In the US, the overlap is considered so important that a whole block is provided as the overlap. It is referred to as "absolute block". This means that there is always a vacant block between trains. It's rather wasteful of space and it reduces capacity but it saves the need to calculate and then build in overlaps for each signal, so it's cheaper. Like a lot of things in life, you get what you pay for.

## Communications Based Train Control

Much of the development effort in the area of new signaling systems is concentrated on removing reliance on track-based equipment. Generically, this type of technology is referred to as Communications Based Train Control (CBTC) and is broadly similar to ETCS Levels 2 or even 3 in as much as conventional track circuits are not necessarily required for train detection. Where trains do not report their own location using tachometers, Doppler radar and balises, detection can be provided additionally by combinations of track circuits, or by means of Global Positioning by Satellite

(GPS) or Global Navigation Satellite Systems (GNSS) as they are now called. Some systems reduce or eliminate the need for fixed track based systems like track circuits and axle counters and rely instead on a combination of train borne speed and location data together with balises and GNSS.

Some CBTC systems use radio-based data bearers to transmit train locations to the central control computers. These are now being used on some metros and people-mover systems like the Las Vegas Monorail. However, some track-based data is still usually considered essential in locations where track separation for parallel tracks and turnout (point) locking is required.

It should be noted that some of the systems using satellite links for train separation were designed in the first instance for areas of "dark territory" where line-side signals are not provided and trains are controlled by train orders and track warrants. The CBTC systems are used to increase throughput and crossing point efficiency whilst not reducing safety.

There is value in considering such systems for use in the UK, particularly where remote routes with low train frequency might have to be closed without a reduction in infrastructure replacement and maintenance costs. GNSS based systems can also be used for passenger information both on trains and at stations.

A further possibility for the use of GNSS is to trigger the warning on the approach to a level crossing with a constant-time lapse regardless of the speed of the train. In a similar way, track workers could be alerted to the approach of a train within a known fixed time.

## Distance To Go – Radio (DTG-R)

The Westinghouse Rail Systems (now Siemens) DTG-R (Distance To Go - Radio) system installed on the London Underground's Victoria Line is the first instance of a full ATP application using radio-based data bearers in the UK. The system is based on fixed block train detection with vital radio data transmission between each interlocking area and the train. It operates in conjunction with ATO and replaced the induction based transmission system originally used on the line.

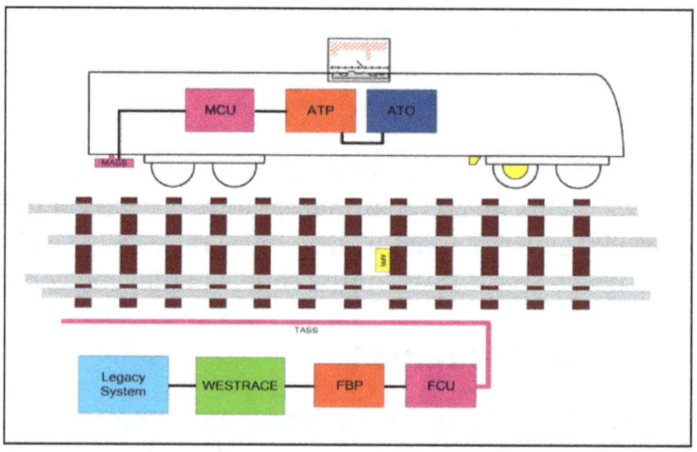


Schematic diagram showing the DTG-R system introduced on the Victoria line in London in 2009. The track to train communication system is radio using the Fixed Communications Unit (FCU) trackside and the on-board Mobile Communications Unit on the train. Transmission is through the trackside antenna (TASS) and the aerial on the train (MASS). The system was overlaid on the legacy system and interlocking data generated by the Westrace system passed to a Fixed Block Processor (FBP). Updates to the on-board map are provided by the Absolute Position References mounted on the track at regular intervals.

The line is divided into what Westinghouse engineers referred to as "interlockings" even where only plain line is provided. An interlocking usually covers a station and the tracks halfway to the next station. Block occupancy data is collected by the Westrace interlocking and passed to a Fixed Block Processor, which converts the data into a radio message. The radio message transmits the block occupancy and proceed status to the train. Geographical data on the line profile and Permanent Speed restrictions (PSRs) is carried on the train. Train positioning is updated by track-mounted balises called Absolute Position References (APRs).

Trains operate under ATO but are capable of operating under full ATP protection with manual driving. The ATO system is similar to the existing Central Line installation using loops known as Platform ATO Communication (PAC) loops. The failure mode restricts manual driving on sight to 10 mile/h.

## Design, Delivery, Migration and Integration

There are a considerable number of technical and operational problems to be overcome before installation of ETCS compatible ATP and migration towards full operation of ATP on the running railway can be achieved. Promoters of any future legislation must be aware of these and of what can be done realistically before any new regulations are introduced. The issues listed below will affect the credibility of those affected as well as the regulators themselves.

- System and component approvals;
- Line capacity;
- The wide variety of existing rolling stock to be fitted with new systems and the effort involved in the physical installation;
- The interfaces between on-board systems and new ATP equipment;
- The interfaces between new ATP equipment and existing signaling and control equipment;
- Mixed system operation;
- New operating rules;
- Conflicts between manufacturers over intellectual property rights;
- Reliability and maintainability;
- Failure modes and degraded operations;
- Interfaces between railway systems (e.g., high-speed and community lines, community lines and metros); Phasing of migration;
- Signal engineering skills shortages.

## Line Capacity

It is a widely held view in the industry that the introduction of ATP will reduce line capacity. This is not necessarily the case in all areas and there are positive benefits from the use of ETCS Level 2/3 in terms of train throughput.

In simple terms, a train's speed on the approach to the block will have to conform to the safe braking distance profile. However, there is no reason why ECTS Level 2 should not be capable of supporting the existing capacity and, indeed, improving on it using shorter fixed blocks. As it does not require fixed blocks, Level 3 will improve capacity in most areas.

## Retrofitting ATP to Rolling Stock

In respect of retrofitting ATP equipment to existing rolling stock, the lessons of the fitting of BR-ATP to the HST Class 43 First Great Western fleet and the Chiltern Railways Class 165/168 DMU fleet should be clearly understood and widely disseminated to operators, suppliers and regulators. These lessons resulted in the need for some time-consuming and expensive measures:

- Overcome underestimated fitting requirements;
- Rewire parts of rolling stock;
- Overcome drivers' resistance;
- Increase reliability;
- Rectify intermittent failures in an intermittent transmission system;
- Reduce equipment self-test and boot-up time;
- Increase maintenance requirements;
- Increase and repeat training;
- Manage TSR and ESR implementation;
- Manage adhesion issues.

It is recommended that a group with a mandate to address lessons learned by both UK and overseas organisations concerned with the design, manufacturing, testing, commissioning, approval and operation of ETCS should be set up and maintained to act as a reservoir of knowledge and experience with the objective of reducing the problems associated with the implementation of new and enhanced ATP systems.

There will doubtless be some rolling stock whose age precludes a cost-effective case for fitting new ETCS equipment and which will continue to operate with legacy equipment.

## Skills Shortages

Any railway signaling developments proposed for the next 10 years will depend on the availability of suitable engineering and installation skills. A large proportion of the relevant UK skills base will be absorbed by the London Underground PPP contracts that cover the re-signaling of over 300 route-km of London Underground lines in the period 2006 to 2017. The equipment to be supplied is novel, ATP based throughout, with complex performance requirements and is still in development. Existing skills shortages in the UK are forcing the contracting companies to source part of the project staff from overseas. Modernisation of the UK main line signaling infrastructure with ETCS based equipment will require similar skills. Shortages, if not addressed immediately, are likely to slow down all replacement programmes.

## Operating Rules

Adapted and new operating rules will need to be written for the new train protection systems introduced to the railway network. Amongst other issues, these rules will have to cover:

- Normal operations, e.g., driving with or without line-side signals under ETCS Levels 2 and 3;
- Failure modes and required actions, e.g., loss of radio contact, failure of on-board equipment;
- Protection of disabled trains;
- Possessions of the track by the maintainer;
- Starting and stabling of trains;
- Movement of trains within station limits;
- Shunting.

Many of these processes will have to change significantly compared with existing practices and will, during the migration periods, have to be run in parallel with them. These circumstances could cause serious delays to train operations or present significant safety risks and should be included in any assessments carried out to determine the methods for introducing ATP.

In particular, the man-machine interface must not be forgotten. The success and acceptance of ATP systems will depend to a substantial extent on drivers' ability to understand and act upon the normal operational modes and failure modes possible with complex ETCS and similar ATP systems. Areas for further exploration concern the man-machine interface, the requirements for revoking authority (replacing signals to danger) and driver supervised movements.

## Roll Back Detection

To provide complete protection, any ATP system adopted will have to incorporate some form of roll-away detection, including roll-back. If this is to be fitted to existing rolling stock, further interface issues will have to be resolved.

## TSRs and ESRs

In any ATP system, the requirement to impose temporary or emergency speed restrictions (TSRs and ESRs) must be considered. In theory, this would seem simply a matter of placing a passive balise on the approach to the speed restriction with its required speed suitably pre-programmed.

However, the issue of gradient effects and how this would be calculated and programmed into the balise must be considered. The quality control of such programming is vital and suitable methodologies must be agreed and monitored. In any case, there is a need to stipulate a standard default mode for temporary balises.

## Migration

A range of complex technical and operational issues will need to be considered as part of the development of migration plans for each area to be upgraded to ATP. These will include:

- Technical interfaces;
- Operational changes;
- Electro-magnetic compatibility;
- Integration with existing systems;
- Retrofitting to existing rolling stock.

Looking at the existing installations of signaling equipment, both at the line side and on rolling stock, it is likely that new ETCS equipment will have to operate alongside existing systems for at least 30 years.

## Degraded Modes: Effects of Failures on Safety and Operations

There are a considerable number of issues related to the effects of failures and the resulting degraded modes of operation. A careful balance must be found between ensuring train movements can continue whilst risk is minimised when operating trains under degraded modes of ECTS and other advanced forms of train protection.

Degraded modes will include the initiation of a brake application by the equipment because of driver error as well as equipment malfunction. A clear distinction should be made between the two but it may be necessary to impose operating rules that assume driver error even in the event of equipment failure. Detailed studies and consultation with regulators and operators are required to ensure the correct rules are understood and agreed prior to equipment installation.

A range of speed limit values for a train protection system, e.g., maximum speeds for different blocks, supervision and intervention speeds, a brake release speed, a shunt speed, a non-supervised speed, a rollback speed, a reverse speed, an on-sight speed etc., will lead to a complex matrix of controls and rules which will lead to poor reliability and possible driver or maintainer confusion.

Basic speed limits will have to be imposed in the event of equipment failure but these should be limited in number and should cover reversing moves, shunting on ATP lines, coupling and uncoupling and recovery modes where possible.

## Degraded Adhesion

All ATP systems fitted to trains operating over open areas of railway infrastructure will be subject to the effects of degraded adhesion from time to time. This is a particular problem in the temperate UK climate and has led to a number of attempts to mitigate the effects on train braking.

A number of strategies have been adopted, ranging from reliance on techniques such as "defensive driving", to sanding and to operating in protected manual mode on ATO metro lines. The Ebicab system allows drivers to input a "poor adhesion" rate when setting up the train for initialisation prior to entering service. Swiss railways use a special code of ETCS to achieve the same objective. This issue must be addressed in respect of both technology and operating rules when introducing ATP.

## Back-up Systems

The use of GSM-R radio for vital data transmission is still regarded throughout the railway industry as either "not fully proven" or "insufficiently reliable" for permanent, full-time operation. Railway administrations, both national and commercial, presently regard the use of radio transmission without a fixed block, visual line side signal back-up system as unacceptable from a reliability perspective. The questions raised about the reliability of GSM-R or GNSS technology and its availability might be answered by the use of two-tier systems or independent, parallel systems being used.

One of the main advantages of ETCS Levels 2/3 is that line-side signals can be removed and that a considerable rationalisation of line-side equipment can be achieved. This is expected to reduce installation and maintenance costs and to improve the life cycle costs. If back-up systems are retained, most of these benefits will be lost and costs will probably rise. This is likely to lead to failure to secure a business case for the viability of ETCS Levels 2/3 installations under existing cost/benefit analysis conditions. To overcome this problem, the reliability of the radio system must be improved and the methods for recovery from failures must be designed carefully to minimise loss of capacity.

In the meantime, experience with the Swiss trial of ETCS Level 2 shows that back-up line-side signaling system should be retained at least during the implementation and stabilisation phases of a project.

## ATP Code Transmission

ATP signaling codes contained in the track circuits are transmitted to the train. They are detected by pick-up antennae (usually two) mounted on the leading end of the train under the driving cab. This data is passed to an on-board decoding and safety processor. The permitted speed is checked against the actual speed and, if the permitted speed is exceeded, a brake application is initiated. In the more modern systems, distance-to-go data will be transmitted to the train as well. The data is also sent to a display in the cab which allows the driver of a manually driven train to respond and drive the train within the permitted speed range.

At the trackside, the signal aspects of the sections ahead are monitored and passed to the code generator for each block. The code generator sends the appropriate codes to the track circuit. The code is detected by the antennae on the train and passed to the on-board computer. the computer will check the actual speed of the train with the speed required by the code and will cause a brake application if the train speed is too high.

## Beacon Transmission

In the examples so far, the ATP data from the track to the train is transmitted by using coded track circuits passing through the running rails. It is known as the "continuous" transmission system because data is passing to the train all the time. However, it does have its limitations. There are transmission losses over longer blocks and this reduces the effective length of a track circuit to about 350 metres. The equipment is also expensive and vulnerable to bad weather, electronic interference, damage, vandalism and theft. To overcome some of these drawbacks, a solution using intermittent transmission of data has been introduced. It uses electronic beacons placed at intervals along the track.

In the best known system, originally developed by Ericsson in Sweden and formerly marketed by Adtranz (now Bombardier), there are usually two beacons, a location beacon to tell the train where it is and a signaling beacon to give the status of the sections ahead. The beacons are sometimes referred to as "balises" after the French. Data processing and the other ATP functions are similar to the continuous transmission system.

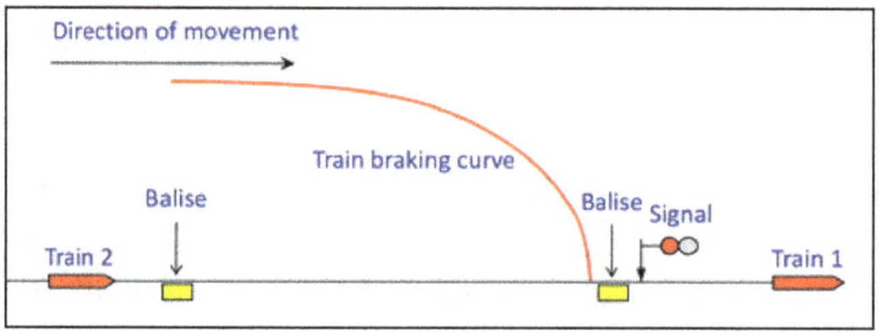


In the diagram, the first balise (beacon) for the red signal gives the approaching Train 2 room to stop. Train 2 will get its stopping command here so that it stops before it reaches the signal. The second beacon causes an emergency stop if the train attempts to pass while the signal is still at danger.

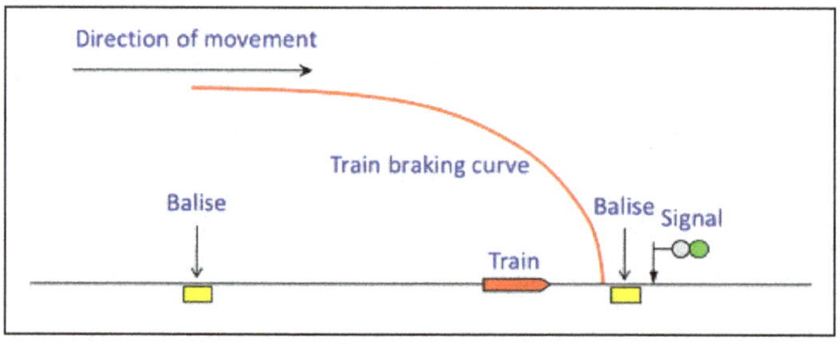


A disadvantage of the beacon system is that once a train has received a message indicating a reduced speed or stop, it will retain that message until it has passed another beacon or has stopped. This means that if the block ahead is cleared before Train 2 reaches its stopping point and the signal changes to green, the train will still have the stop message and will stop, even though it doesn't have to. Why, might you ask, can't the driver cancel the stop message like he does when the train has stopped and the signal changes to green? If he could cancel the stop message while the train was moving, the system would be no better than the AWS with its cancel button. ATP is "vital" or "fail-safe" and must not allow human intervention to reduce its effectiveness.

In reality, it will not move even then, since it requires the driver to reset the system to allow the train to be restarted. For this reason, this type of ATP is normally used on manually driven systems.

## Intermittent Updates

To avoid the situation of an unnecessary stop, an intermediate beacon maybe provided or, an infill loop. This updates the train as it approaches the stopping point and will revoke the stop command if the signal has cleared. More than one intermediate beacon can be provided if necessary.

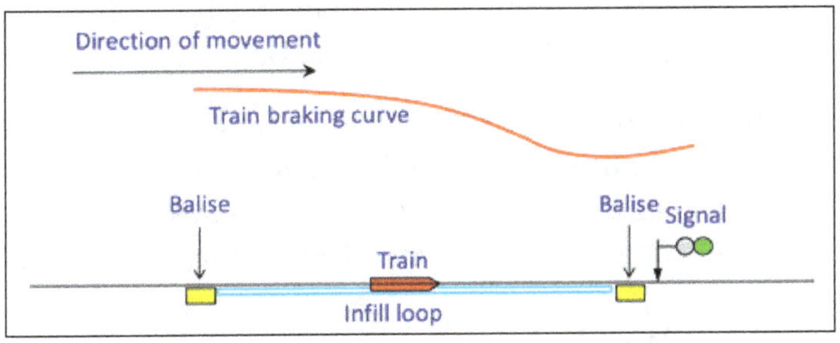


## Moving Block

As signaling technology has developed, there have been many refinements to the block system but, in recent years, the emphasis has been on attempts to get rid of fixed blocks altogether. Getting rid of fixed blocks has the advantage that you can vary the distances between trains according to their actual speed and according their speeds in relation to each other. It's rather like applying the freeway rules for speed separation - you don't need to be a full speed braking distance from the car in front because he won't stop dead. If you are moving at the same speed as he is, you could, in theory, travel immediately behind him and, when he brakes, you do. If you allow a few metres for reaction time to his brake lights and variations in braking performance, it works well. Although it only needs a few spectacular collisions on the freeways to disprove the theory for road traffic, in the more regulated world of the railway, although it could not be applied without a full safe braking distance between trains, it has possibilities.

As long as each train is travelling at the same speed as the one in front and they all have the same braking capabilities, they can, in theory, run as close together as a few metres. Just allow some room for reaction time and small errors and trains could run as close together as 50 metres at 50 km/h. Well, that's OK in theory but, in practice, it's a different matter and, as yet, no one has taken moving block design this far and they are unlikely to do so in the near future. The recent ICE high speed accident in Germany where a train derailed, struck a bridge and stopped very quickly, effectively negates the safety value of the theoretical moving block system described above. This means that it is essential to maintain a safe braking distance between trains at all times.

What is worth doing, is making the the block locations and lengths consistent with train location and speed, i.e. making them movable rather than fixed. This flexibility requires radio transmission, sometimes called Communications Based Train Control (CBTC) or Transmission Based Signaling (TBS) rather than track circuit transmission, to detect the location, speed and direction of trains and to tell trains their permitted operating speed.

## Moving Block and Radio Transmission

On a moving block equipped railway, the line is usually divided into areas or regions, each area under the control of a computer and each with its own radio transmission system. Each train transmits its identity, location, direction and speed to the area computer which makes the necessary calculations for safe train separation and transmits this to the following train.

The radio link between each train and the area computer is continuous so the computer knows the location of all the trains in its area all the time. It transmits to each train the location of the train in front and gives it a braking curve to enable it to stop before it reaches that train. In effect, it is a dynamic distance-to-go system. This is Communications Based Train Control (CBTC).

One fixed block feature has been retained - the requirement for a full speed braking distance between trains. This ensures that, if the radio link is lost, the latest data retained on board the following train will cause it to stop before it reaches the preceding train. The freeway style vision of two trains moving at 50 km/h with 50 metres between them is a step too far into virtual reality for most operators.

## Moving Block - Location Updates

Trains in a moving block system report their position continuously to the area computer by means of the train to wayside radio. Each train also confirms its own position on the ground from beacons, located at intervals along the track, which recalibrate the train's position compared with the on-board, computerised line map.

Transferring a train from one area to another is also carried out by using the radio links and, additionally by a link between the two adjacent area computers. The areas overlap each other so, when a train first reaches the boundary of a new area, the computer of the first area contacts the computer of the second area and alerts it to listen for the new train's signal. It also tells the train to change its radio codes to match the new area. When the new area picks up the ID of the train it acknowledges the handover from the first area and the transfer is complete.

Another version of the moving block system has the location computers on the trains. Each train knows where it is in relation to all the other trains and sets its safe speeds using this data. It has the advantage that there is less wayside equipment required than with the off-train system but the amount of transmissions is much greater.

## An Early Moving Block System

One system which claims the distinction of being the first moving block system is that marketed under the name Seltrac by Alcatel. It is used in Canada and on the Docklands Light Railway in London. It has the ingredients of moving transmission of data, but the transmission medium is the track-mounted induction loops which are laid between the rails and which cross every 25 metres to allow trains to verify their position. Data is passed between the vehicle on-board computer (VOBC) and the vehicle control centre (VCC) through the loops. The VCC controls the speed of Train 2 by checking the position of Train 1 and calculating its safe braking curve. The Seltrac system requires no driver, as it is fully automatic. In case of a system failure where a train has to be manually driven, it has axle counters to verify the position of a train not under the control of the loops. Perhaps its biggest drawback is the need for continuous cables to be laid within the tracks, expensive to install and open to damage during track maintenance.

The principle difference between this system and the more modern ones being marketed today is that Seltrac uses electro-magnetic transmission of data requiring track cables, whereas radio based systems only require aerials. Seltrac is upgrading their design to use radio based transmission.

### Moving Block - Why Do We Need It?

Railway signaling has traditionally required a large amount of expensive hardware to be distributed all along a route which is exposed to variable climatic conditions, wear, vandalism, theft and heavy usage. Because of the widely spaced distribution, maintenance is expensive and often restricted to times when trains are not running. Failures are difficult to locate and difficult to reach. On metros, access is further restricted where there are tunnels and elevated sections. For these reasons, railway operators have been trying to reduce the wayside signaling equipment and so reduce maintenance costs. Reduced wayside equipment can also lead to reduced installation costs. Moving block requires less wayside equipment than fixed block systems.

There is another goal much sought after by operators - greater capacity. A norm for most metro lines is 30 trains per hour (tph) or a two-minute headway. It is debatable whether much improvement on this is possible for a high capacity system, since the major losses of line capacity occur because of station stops and terminal operations. Heavily used metro lines, like those in Hong Kong, trying for a greater capacity than 30 trains per hour, will struggle to keep dwell times below 40-50 seconds at peak times. This will push the headway to two minutes or longer, regardless of the signaling system used. Similar problems exist at terminals where crossover clearance times are critical. Moving block signaling cannot provide much improvement. Shorter headways can, however, be achieved on systems where trains are shorter, speeds lower and the passenger levels smaller. In some places a 95 second headway can be achieved on systems like Docklands and certain sections of the Paris Metro.

Also, for underground lines, modern ventilation and smoke control systems will require train separation of 2-300 metres to allow air circulation at critical times. If moving block signaling allows 50 metre separation, some very expensive additional ventilation arrangements might be necessary. This may reduce the benefits of moving block.

The real prize which could be won by an operator using moving block is reduced wayside equipment and reduced maintenance costs. Better reliability and quicker fault location is also possible with moving block technology. If radio based transmission is included, an all-round improvement can be achieved.

One other factor to be noted is that many operators specifying moving block technology also ask for fixed block track circuits to serve as a back up and for broken rail detection. Track circuits are also still required for junctions.

## Anti-collision Device

The anti-collision device is a form of automatic train protection used on Indian Railways. The ACD Network is a train-collision prevention system invented by Rajaram Bojji and patented by Konkan Railway Corporation Limited, a public-sector undertaking of the Ministry of Railways, government of India. ACDs rely on GPS satellites for position updates. They exchange information through radio frequency transmissions to automatically brake and prevent collisions.

Loco ACDs brake to reduce the train speed to 15 km/h if on approach. They receive messages from each other ACDs on adjacent tracks. If inbound ACDs of other trains read 'train parted', the trains decelerate until they have stopped to prevent dangerous side collisions that can occur when adjacent tracks have been damaged. The experimental trial use of ACDs in the Southern Railway has been successful.

## Level Crossings

When Loco ACDs receive 'Gate Open' transmissions from Gate ACDs provided at non-interlocked level crossings, they brake to decelerate to 30 km/h or an alternative predetermined speed. Gate ACDs at manned and unmanned level crossings also warn passengers with the message 'Train Approach'.

If a Loco ACD receives a manual 'SOS' message from other train bound ACDs or a station ACD that is within three kilometres of its radial range, it applies brakes automatically to bring the train to a stop.

The application of this anti-collision device has been refined to not only prevent midsection collisions but also to prevent their occurrences in station yards. The newly engineered solution is integrated with the signaling systems, interlocking to react appropriately in case collision-like conditions are perceived at the time of reception and dispatch of trains from a station (e.g. while approaching a station). Loco ACDs also give 'Station Approach' alerts to train operators and regulate train speed when they receives information from Station ACDs.

Loco shed ACDs, Track-ID Assigning ACDs and Repeater ACDs strengthen the ACD network.

## Future

Indian Railways have successfully piloted ACDs in the northeast frontier railway, covering 1,736 kilometres (1,079 mi) of its broad gauge route. They are now installing the ACDs on 760 kilometres (470 mi) of the Konkan Railway.

The on-board train protection device, the first device designed by Konkan Railway with their technical partner Kernex Microsystems (I) Ltd, was installed throughout the Indian Railway network.

A new ACD Version-II, now called the Train Collision Avoidance System (TCAS), is under development by The Research Designs and Standards Organisation (RDSO). Unlike ACD, which is more of a distributed system which acts independently, the TCAS will be more of a centralized system where TCAS controls communication between trains and with trains with the TDMA protocol. The TCAS under development is meant to be a vital safety system. TCAS has a deep coupling with the railway signaling system so ACD systems do not depend on the railway signaling system.

## ACD Deficiencies

The ACD system is based on GPS based positioning and track detection. This has inherent problems as with GPS service and course acquisition, the best possible horizontal accuracy is 10 m. This is inadequate for detection of rail tracks separated by a distance of 10–15 feet. ACD does not even have DGPS, differential GPS that gives an accuracy close to 2.5 m, and hence had errors in track detection using their patented Deviation Count Theory that worked in block sections but failed in station sections. The result was erratic braking that disrupted train movements and proved to be ineffective.

The Railway Collision Avoidance System was patented in 2001 by an Indian inventor, Indranil Majumdar from Calcutta. He was awarded the Texas Instruments Analog Design Challenge 2001 for this design and another patent was granted in 2007. The design ruled out GPS as it was a 3rd party US based system. Instead it uses track based sensors similar to RFID or Balise similar to the EuroBalise. The design did not receive much attention as Majumdar had no railway background. However, the design concepts were eventually reflected in the TCAS design first released in 2008.

After seven or eight years of problems with the ACD system, RDSO, Lucknow drafted the Train Collision Avoidance System (TCAS) specifications with amendments. In 2012, the Ver3.1.1 specification was released after joint consultation with companies manufacturing signaling equipment for the Indian Railways. The ACD system used in Indian Railways has inherent problems in station sections due to their design, using GPS for unfeasible track detection.

The High-Level Safety Review Committee at Mumbai on 12-13 January 2012 at the Western Railway HQ was skeptical of ACD effectiveness. They unanimously chose to develop TCAS as an open architecture system without charging royalties unlike the ACD which is proprietary.

TCAS is being developed in India by qualified companies, manufacturing railway safety systems selected by RDSO through an Expression of Interest (EOI). These companies includes Kernex Microsystems, Medha Servo Drives Hyderabad, Invensys Bangalore, Siemens, HBL Power Systems Ltd Hyderabad and others. Indian Train Protection Systems will offer collision avoidance and also many functionalities of the European Train Control System, including prevention of Signal Passing at Danger (SPAD), Movement Authority and Control, Critical Train Data Recorder, advance in cab display of signals, advance alerts and warnings from station sections, uploading of running train data to a Central Train Management System over GSM-GPRS or other cellular networks.

The Indian government selected TCAS for future installation at a cost of 10 L INR per kilometre.

## Emergency Brake

Driver's brake handle in a UK Electric Multiple Unit.

On trains, the expression emergency brake has several meanings:

- The *maximum* brake force available to the engine driver from his/her conventional braking system, usually operated by taking the brake handle to its furthest position, through a gate mechanism, or by pushing a separate plunger in the cab.

- A completely separate mechanism from the conventional braking system, designed to stop the train as quickly as possible.
- A handle or plunger which may be applied by a passenger in an emergency situation, either stopping the train directly or sending an alarm to the driver so that he/she can stop the train.

The emergency brake applies considerably more braking force than the standard full-service brake. The engine driver or motorman will only use the emergency brake as a last resort, since it may cause damage; even with modern wheel slide protection, a train may develop wheel-flats, and the rails themselves can suffer profile damage.

## Possible Consequences of Operation

Putting the engine driver's brake handle into the emergency position may also:

- Shut off traction power.
- On an electric train, drop the pantograph if the power supply is overhead-wire traction current, or raise the pickup shoes where the supply is third rail traction current.
- Disable dynamic braking.
- Prevent the brakes from being released until the train has stopped.
- Connect the brake wires to ground, to prevent a "false-feed" voltage from releasing the brakes.
- Send an emergency radio signal, if the brake handle has been in the emergency position for over 30 seconds.

## Passenger-applied Brakes

Trains often have a facility in each car to enable passengers to apply the brakes in case of emergency. In many modern trains, the driver is able to prevent brake activation when a passenger operates the emergency alarm - an audible warning is sounded, and the driver is then able to talk to the person who activated the alarm on the intercom and see them on an internal CCTV. The driver can press an override button and hold the brakes off whilst he/she chooses a safe place to stop the train. Because of possible serious problems, severe fines and/or imprisonment penalties are in place to stop people from activating the brake without good reason.

## Mechanism on Locomotive-hauled Stock with Brake Pipe

The alarm chain in a passenger coach is designed to create a break in the continuity of the brake pipes (whether vacuum or air brakes), immediately resulting in a loss of brake pressure (or vacuum) and thereby causing the train brakes to be applied. With vacuum brakes, a clappet valve is provided, which is released by the pulling of the alarm chain; with air brakes, there is a similar passenger emergency valve which can vent the brake pipe to the air.

In most locomotives (in addition to a warning lamp or buzzer being sounded) the master controller undergoes auto-regression, with the notches falling to zero rapidly as the locomotive's motive power is switched off. The guard may also notice the loss of brake pressure (although he/she may not know it is due to the pulling of the alarm chain) and is expected to apply their brakes immediately as well. It is possible for a driver to override the alarm-chain pull; this is sometimes done where it is known that miscreants resort to pulling the emergency chain solely to get the train to stop at a point convenient for themselves. However, such an act by the driver (or guard) of deliberately ignoring an indication of alarm-chain pulling is a serious matter.

In recent years locomotives have been fitted with emergency flashers on the roof of the cab, and these flashers are also activated when the brake pipe pressure is lost for any reason other than the driver's application of the brake valve. This alerts drivers of oncoming trains of the possibility of a derailed or parted rake which may foul other tracks (since brake pressure may have been lost for those reasons as well); at the locomotive, it is not possible to tell whether the loss of brake pressure is due to the pulling of the alarm chain.

ACP (alarm chain pulling) also causes a small lever to be released near the emergency brake valve (usually mounted near one end of the coach), which does not retract to its normal position even when the chain is released. This allows the driver (or guard) to find out in which coach the ACP actually occurred. When the coach is isolated, the lever needs to be manually reset. Until this is done, the lamp and buzzer in the locomotive cab are continuously activated. A circuit breaker controls the lamp and alarm bell in the locomotive cab; in cases where defective equipment causes the lamp and bell to go off, the driver can disable them by placing the MCB (miniature circuit breaker) in the "Off" position. Despite the obvious safety hazards, sometimes this is resorted to by drivers when driving trains through sections where spurious ACP incidents are very common.

## Wheel Slide Protection

Wheel slide protection and wheel slip protection are railway terms used to describe automatic systems used to detect and prevent wheel-slide during braking or wheel-slip during acceleration. This is analogous to ABS and traction control systems used on motor vehicles. It is particularly important in slippery rail conditions.

### Sanding

Sanding equipmment on a ČD class 971 driving trailer.

Sanding is one method of reducing wheel slip or slide. Locomotives and Multiple units have sandboxes which can deliver dry sand to the rails in front of the wheels. This may be initiated automatically when the Wheel Slide Protection system senses loss of adhesion, or the driver can operate it manually. Sanding may be connected to a computer system that determines the train's direction and where the sand should be applied: either forward or aft of the trucks. In older locomotives there was a manual lever attached to a valve that had three positions: Off, Forward, and Aft.

## Automatic Control Systems

Wheel Slide Protection (WSP) equipment is generally fitted to passenger trains to manage the behaviour of wheel sets in "low adhesion" (reduced wheel/rail friction) conditions. It is used when braking, and may be considered analogous to anti-lock braking (ABS) for cars. The system can also be used to control (or provide an input to) the traction system to control wheel spin when applying power in low adhesion conditions.

"Low adhesion" at the rail potentially causes damage to wheels and the rails. Typically, low adhesion conditions are associated with environmental causes arising from seasonal leaf fall, or industrial pollution. Occasionally the cause can be another less obvious factor such as light oxidation of the railhead or even swarms of insects.

## Wheel Slide whilst Braking

When a train is braking, the low adhesion manifests as wheel slip where the wheelset is rotating at a lower velocity (speed) than the forward speed of the train. The most extreme example of this is where the wheel stops rotating altogether (wheel slide) while the train is still moving and can result in a "wheel flat" caused by the softer wheel steel being abraded away by the harder rail steel.

However, the wheelset does not need to lock up completely in order for damage to be caused. If the slide is significant, heat can build up in the contact patch between wheel and rail sufficiently to permanently modify the crystalline structure of the wheel's steel. The steel becomes more brittle (martensite) which leads to cavities forming in the wheel. Wheel flats on railway vehicles are very evident by their characteristic "bang-bang" in time with the speed of the train. It is normally necessary to use a wheel lathe to remove a layer of wheel tread caused by a severe flat or cavity, which reduces the life of the wheel and is a major operating cost to the rail industry.

## Wheel Slip whilst Taking Power

In traction, low adhesion may cause a wheelset to accelerate more quickly than the train (wheel spin) to the point where it can damage the traction system or result in damage to the wheel and rail (rail burn).

## Controlled Wheel Slip

WSP is generally fitted as standard to new fleets of multiple units. The primary function of the WSP is to improve the ability of a train to stop in poor adhesion conditions. However, within the rail industry it is also recognised to be valuable in protecting the wheels from damage during sliding when braking or spinning in traction. This improvement is achieved by regulation of the wheelset velocity in a controlled manner so that it maintains a relatively consistent level of slip. The controlled slip has the effect of conditioning the contamination layer on the rail (scrubbing action) thereby improving the level of friction and enhancing the ability of the train to stop. Controlled wheel slip can also have a limited cleaning action on the rail head along the length of a train. This tends to result in the vehicles at the rear having more grip than those at the front.

## Driving Technique

WSP continuously monitors the rotational speed of axles on the loco or train, and intervenes whenever it detects a significant difference on any axle.

If wheelslip occurs whilst the power controller is open, WSP will shut off power to the affected traction motor(s). Despite this most train companies advise their drivers to close the power controller and allow the slipping wheels to stabilize before re-opening the controller on a low setting because control of the train can be achieved more quickly.

However when wheelslide occurs and the WSP releases the brakes on affected axles, drivers are instructed to leave the brake handle alone and let the WSP control the train's braking. This is because the driver is sitting over the leading bogie of the train where wheelslide is usually most severe. This wheelslide will partially clean the railhead and so further down the train the wheels will achieve better adhesion, and thus braking effect.

## 'Low Adhesion' Training

Driving a train under low adhesion conditions takes experience. Failing to recognize and respond correctly to railhead contamination or environmental conditions which cause low adhesion can lead to safety incidents such as a Signal passed at danger, collision or station overrun.

Prior to each 'Leaf fall season' many train companies arrange low adhesion training for their newly qualified drivers. This consists of taking over a section of line during a quiet period. Using lineside markers each driver gets his train up to speed and then makes a full service brake under normal adhesion conditions. The railhead is then treated with a contaminant that has a low coefficient of friction. On the second run the driver will experience the sound and sensation of the WSP activating and operating blowdown valves on the brake cylinders, and the stopping distance will be considerably greater.

Although this only provides an approximation of how a train will behave during low adhesion, it does ensure that the driver can recognize the onset of wheelslide and will know the correct actions to take when it occurs.

## Microprocessor Control

Modern WSP systems are microprocessor controlled and employ two stage valves that permit fine control over the air pressure in the brake cylinders. This is essential to be able to capture and control a sliding wheel and to minimise the amount of air resource used by the WSP. When the brake is applied, the WSP first applies the dynamic brake. If that is not successful it then "blends" the friction and dynamic braking systems. If control is still not established the system reverts to friction braking only where blowdown valves rapidly cycle the air in the brake cylinders. Examples of this kind of equipment are manufactured by Knorr Bremse (EP compact, EP2002) Faiveley Transport (EPAC) and POLI Wabtec (ATHENA).

### Manufacturers

Manufacturers of WSP equipment include Faiveley Transport, Knorr-Bremse, Wabtec, DAKO, KES & Co GmbH, Mitsubishi, Siemens, Selectron Systems AG and ABB.

### Testing

Demonstrating the improvement provided by a WSP system is very difficult as the naturally occurring low adhesion condition occurring at the rail can be difficult to re-create in a test track environment.

### Track Testing

For track testing, a detergent based solution has historically been used to provide low adhesion test conditions. European and international standards often refer to this test method (BS-EN 15595, UIC 541-05). In the UK, British Rail Research adopted two approaches including a laboratory simulation method to all WSP approvals from around 1992, and track testing using carefully conditioned paper tape adhered to the railhead. The paper tape method used in the UK is believed to offer a realistic representation of the challenging very low adhesion conditions encountered during the autumn leaf fall. With an increase in the privatisation of railways in Europe, track testing has become very expensive to organise and to conduct. As a consequence, simulation based testing is rapidly becoming more popular with WSP manufacturers and national bodies.

### Simulation Testing

Simulation testing employs a computer representation of the train and the track conditions, and signals are provided to the WSP system that effectively deceive it into thinking it is fitted to an actual train. Most of the WSP manufacturers have some simulation capability and there are also facilities available from national bodies or independent testing facilities such as Deutsche Bahn (DB Germany), Ferrovie dello Stato (FS Italy), and DeltaRail Group (formally BR research) (UK and Ireland).

## Railway Slide Fence

Part of a railway signaling system, a slide fence is a fence whose purpose is to prevent trains from being derailed by rock slides in mountainous areas where rock slides may occur without warning. The fence is designed to be displaced by a rock slide, causing the signaling system to display a stop aspect on nearby signals. As an alternative, a structural fence is designed to physically stop falling rocks from reaching the tracks.

### Operation

### Mechanical

The mechanical slide fence consists of a series of tensioned wires strung about 10 inches (25 cm) apart on poles. When a rock slide occurs, the rock breaks one or more of the wires. When a wire breaks, heavy weights attached to either end will fall. This mechanically triggers the protecting signals to the 'danger' position.

## Electrical

There are two types of electrical slide fences in operation.

One type of electrical slide fence consists of a series of parallel conductive wires strung about 8 inches (20 cm) apart on poles that create a fence parallel to the rails.

This creates an electrical circuit that is monitored by signaling equipment. In normal operation, the electric current in the fence wires causes a relay to energize, indicating that the fence is intact. When a rock slide occurs, the rock breaks one or more of the wires, interrupting the current. This causes the relay to become de-energized, indicating that a rock slide has occurred. A contact of the relay is typically used to prevent the approaching signal from displaying a proceed aspect if the fence has been broken. Restoring normal operation requires splicing the broken fence wires back together.

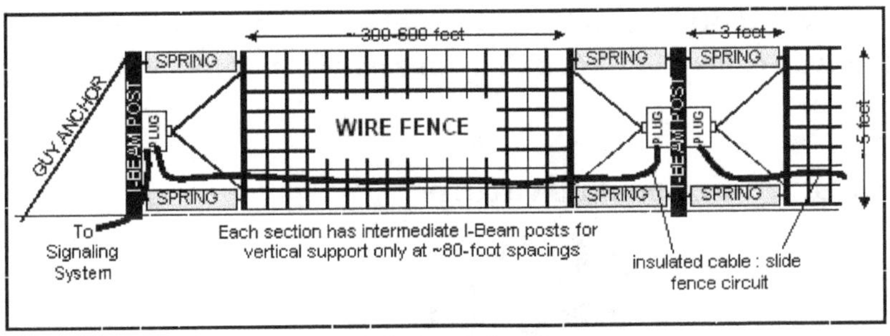


Slide fence detail

Another type of slide fence is similar except that the wires do not have to break and is easier to maintain and reset. The slide fence consists of a series of fence sections, which shows a typical installation. Each fence section is held in place by strong springs. At each end of a fence section is an electro-mechanical plug. The plugs maintain a complete electrical circuit that is monitored by signaling equipment. In normal operation, the current through the plugs causes a relay to energize, indicating that the fence is in place. When a rock slide occurs, the fence moves laterally, causing the plug to be removed, breaking the circuit. This causes the relay to become de-energized, indicating that a rock slide has occurred. A contact of the relay is typically used to prevent the approaching signal from displaying a proceed aspect if the fence has been broken. Restoring normal operation requires re-inserting the plugs that were dislodged by the slide.

## Structural Fence

The structural fence is a physical barrier designed to stop falling rocks from reaching the tracks. Several construction methods are used, including: steel I-beams, wooden barriers, galvanized fencing, and netting directly against the rock.

## Usage

A slide fence is typically found in mountainous areas in a rock cut area, where rocks may fall on the track and present a danger to passing trains. The length of the fence may range from 100 feet (30 meters) to several miles (kilometers), depending on the length of the rock cut and the area being protected. The slide fence is usually located on the uphill side of the track in the slide area.

A BNSF oil train is eastbound in Bad Rock Canyon where railway slide fences are installed.

Select examples of railroad slide fences:

- On the Union Pacific Railroad's Moffat Subdivision, trains pass through three major canyons — Byers Canyon, Gore Canyon, and Glenwood Canyon, all three of which feature rock fences.
- Stretches of the BNSF Railway's Hi-Line, such as on the western approach to Marias Pass, and a small canyon called Bad Rock Canyon east of Columbia Falls, Montana, are equipped with rock fences.

## Consequences of Slide Fence Activation

Hand-written slide protection placard on a PRR Pedestal type signal at CP-SOUTH FERRY indicates to engine crews that this signal is connected to a slide detector.

When a train approaches a slide fence area, and the signal displays a stop aspect, the train is not permitted to proceed normally, because a rock slide has occurred. However, the rock slide may not prevent the safe passage of the train. For example, a large rock may have fallen off the cliff, broken through the slide fence, and continued to fall away from the track. After stopping, the train may obtain radio permission from the dispatch center to proceed slowly, watching for a dangerous rock slide. If the train is able to successfully pass through the slide area (that is, there is no danger), it may then be allowed to proceed normally.

In North America slide fences are typically connected in such a way as to shunt the track circuit when activated. This causes signals on either side of the slide fence to display a restricting indication, requiring trains to travel at a speed enabling them to stop within one half the range of vision. On lines formerly operated by the Pennsylvania Railroad signals connected to a slide detector have an 'SP' placard, reminding engineers to watch for slides when governed by a restrictive speed signal.

Once the slide fence has been activated (even if in error), all trains are affected until the slide fence is repaired by maintenance personnel. This may result in several hours of delay in train service.

### Alternatives

Several alternative technologies have been tried to solve the rock slide problem, including:

- Acoustic sensing
- Electromagnetic sensing
- Seismic sensing
- Visual sensing, using cameras

## Defect Detector

A defect detector is a device used on railroads to detect axle and signal problems in passing trains. The detectors are normally integrated into the tracks and often include sensors to detect several different kinds of problems that could occur. Defect detectors were one invention which enabled American railroads to eliminate the caboose at the rear of the train, as well as various station agents stationed along the route to detect unsafe conditions. The use of defect detectors has since spread to other overseas railroads.

### Sensors

Hotbox Detector.

The sensors installed at defect detector locations can include and are explained.

### Hotbox

Hot boxes or Hot Bearing Detectors are used to measure the temperature of the journal bearings of a train. They typically consist of two infrared eyes on each side of the tracks looking up at the train's bearings. They register the radiation from every journal that passes over them. If a bearing reaches the maximum temperature for safe travel, the detector will flag and count it as a defect.

## Dragging Equipment Detectors

A column of cones sits across the whole width of the railroad (just like a cross tie) attached to a switch. Anything dragging from the train will hit this cone, thus pushing it back, thus breaking a contact. It then returns to its normal position to prepare for anything else that might be dragging under the train. The detector will register this action and flag it as a defect. Brittle bars are still used elsewhere, but still have to be repaired. Over time, dragging equipment detector's metal flaps need to be replaced because of extensive damage to them. Single use systems typically involve a frangible engagement bar or a stainless steel wire/braid strung between the rails and typically outside the rails as well, fastened to the sleepers. If the bar or braid is hit by something it breaks, and the circuit break alerts that there is a dragging item. Auto-resetting systems typically involve a pivot pin system to allow the target to reset itself after a hit.

## High Car or Shifted Load Detectors

Infrared beams are placed horizontally above the track (high car) or vertically next to the track (shifted load). Anything that breaks the beam will be counted as a defect. A high car detector is placed anywhere an excess height car could be routed onto a low clearance line. A shifted load detector is mainly found on railroads where double-stack trains are prevalent as the containers may become misaligned and present a hazard to bridge trusses or tunnel walls.

## Wheel Impact Detectors

Wheel sensors along the tracks feel for flat spots on the train's wheels. Any flat wheel that becomes too dangerous to travel on (a big flat spot on the train wheel) will be counted as a defect. Typically, these systems utilize accelerometers, strain gauges, fiber optic methods, or the very latest (WIPD) wheel impact phase detector. A wheel impact load detector (WILD) measures impacts, but does not normalize these impact measurements against anything: simply the impact reading. They do not attempt to cater for differences in sprung mass, as they are measuring wheel defect impact rather than impact load. Therefore, the same wheel defect will register a much larger impact when a wagon is loaded, versus when it is empty. A wheel condition monitoring detector monitors the condition of the wheel independent of sprung mass – independent of load. They do this by subtracting the wheel mass to get the normalized impact value. These systems are therefore typically better at detecting smaller defects with greater resolution.

## Sliding/Hot Wheel Detector

Typically a length of side looking infrared detectors that can detect if a wheel has locked up and is sliding along the track or has had the brakes lock up causing the entire wheel to heat up.

## Vehicle Imbalance Detectors (Side-to-Side and End-to-End)

These detectors are typically a crude weighbridge and/or WILD system, as they are only concerned with measuring weight differentials. They do not have to be as accurate as proper weighbridge or

WILD systems, but just accurate enough to be able to average the weight of bogies during a train pass to calculate the relative balance of wagons, to establish if one rail is loaded unacceptably greater (in percentage) than the other. This is typically not performed on empty wagons because of the significant percentage imbalances that can be caused by fluctuations in weight due to bogie tracking geometry or hunting issues, which in terms of weight differentials are relatively accentuated compared to when a wagon is loaded.

## Clearance Gauge Monitors

These detectors can use a variety of sensors (video, laser, infrared), but typically they are a safety curtain arrangement: a gantry over the rail with sensors to detect anything outside the Clearance Gauge. They therefore check the envelope of safety and alert if anything is detected outside of it.

## Low Hose Detectors (Air Hose)

These systems are different than dragging equipment detectors, which are looking for anything dragging from the train connecting with the sleepers. Low Hose detectors are looking specifically between two wagons, to measure the droop of the brake hoses. The brake hoses need a bit of droop, but too much and they can hit or dislodge. Typically, these systems employ a vertical bar installed on one side of the track, with infrared sensors pointed across the track at a mate bar with receivers installed pointed back at the infrared beams. When a train goes past, the infrared sensors are looking in the coupling area specifically, and will alert if anything is detected in the area that is lower than acceptable.

## Video/Image Inspection Systems

These systems rely on an array of video devices in various locations between the rails and either side of the track, looking for particular bogie components (such as brake beam, springs, friction wedges, etc) and this data is then put through image analysis to determine if there are maintenance issues.

## Wheel Profile Systems

These are laser arrays installed between or in place of a sleeper. They point up at the wheel profile and measure the profile shape specifically, taking various measurements of angles and lengths. The systems also give differential measurements for the two wheels of a wheelset.

## Bogie Performance Detectors

Bogie Performance Detector - TBOGI system.

Bogie performance detectors monitor bogie tracking geometry, and hunting (instability) behaviour. Bogie tracking geometry includes tracking position and angle of attack on a per-axle basis, as well as rotation, shift, inter-axle misalignment, and tracking error on a per-bogie basis. Bogie performance detectors can provide early detection of bogie defects, and early warning of derailment risks through flange climb or rail break. Bogie performance detectors most often use optical methods, and are installed adjacent to the track with wheel sensors clamped to the rails.

### Rail Break Monitors

These detectors either inject frequencies into the rail, or rely on existing track circuits, to detect rail breaks. These detectors are most commonly used by high speed networks.

### Rail Temperature Monitors

These detectors are a mesh of small detectors installed against a particular stretch of rail. The detectors are typically a mix of temperature sensors and strain gauges (so measure in degrees Celsius and kilonewtons force). They measure temperature and stress/tension on the rail, to make sure these measures do not go outside the bounds of structural integrity. These systems are initially calibrated to a particular 'neutral state' but measuring the rail with neutral compression (no pulling or pushing) and an agreed neutral ambient temperature. The system then measures the rail to detect if these parameters get too far from neutral state, and will alert if the rail is approaching a breach of structural integrity.

### Ground Based Noise (GBN)

This is a combination of accelerometers on the track, and another accelerometer installed in the bedrock next to the track. The correlation of these measurements indicates how much track-born noise is propagating through the ballast into the bedrock strata. This correlates directly with the noise levels experienced by surrounding area. These systems are commonly installed near or inside tunnels.

### Brake Block Detectors

These are distinct from generic video imaging systems, as they are dedicated to specifically imaging the band above the rail that contains the brake shoe. The system identifies the back plate of the brake shoe, and then works out how much brake block is left in millimetres. If the system detects no brake block beyond the back plate, the system typically identifies the pad as missing.

Sensors for the wheel condition monitor.

### Wide-load Detectors

A bridge spans the railroad with two laser beams that shine down on each side of the passing train. Anything that cuts through the beam will be counted as a defect. This sensor may also be integrated into the high car detector.

## Positive Train Control

Positive train control (PTC) is a system of functional requirements for monitoring and controlling train movements and is a type of train protection system. The term stems from control engineering. The train is only allowed to move in case of *positive* movement allowance. It generally improves the safety of railway traffic.

Train protection systems are used to control traffic movement by technical means. They are especially needed in cases of high speed transportation, dense traffic with short succession of trains and mixed type traffic at widely differing speeds. Train protection systems were in practical testing at least since the beginning of the 1930s in Europe. Stopping a running train is the main goal of any *train protection system*. This is most easily done with stop order, and without a special order the vehicle is allowed to run. A typical representative for this "negative train control" is Indusi. In contrast to this 'easy moving', a PTC restricts the train movement to an explicit allowance; movement is halted upon invalidation.

The main concept of PTC (as defined for North American Class I freight railroads) is that the train receives information about its location and where it is allowed to safely travel, also known as movement authorities. Equipment on board the train then enforces this, preventing unsafe movement. PTC systems may work in either dark territory or signaled territory, and may use GPS navigation to track train movements. Various other benefits are sometimes associated with PTC such as increased fuel efficiency or locomotive diagnostics; these are benefits that can be achieved by having a wireless data system to transmit the information, whether it be for PTC or other applications.

The Federal Railroad Administration (FRA) has listed among its goals, "To deploy the Nationwide Differential Global Positioning System (NDGPS) as a nationwide, uniform, and continuous positioning system, suitable for train control." The U.S. freight rail industry says that at the end of 2018, the nation's largest freight railroads were operating PTC across 83.2 percent of the required route miles. The American Railway Engineering and Maintenance-of-Way Association (AREMA) describes Positive Train Control as having these primary characteristics:

- Train separation or collision avoidance
- Line speed enforcement
- Temporary speed restrictions
- Rail worker wayside safety

## Basic Operation

A typical PTC system involves two basic components:

- Speed display and control unit on the locomotive.
- A method to dynamically inform the speed control unit of changing track or signal conditions.

Optionally, three additional components may exist:

- An on-board navigation system and track profile database to enforce fixed speed limits.
- A bi-directional data link to inform signaling equipment of the train's presence.
- Centralized systems to directly issue movement authorities to trains.

## PTC Infrastructure

There are two main PTC implementation methods currently being developed. The first makes use of fixed signaling infrastructure such as coded track circuits and wireless transponders to communicate with the onboard speed control unit. The other makes use of wireless data radios spread out along the line to transmit the dynamic information. The wireless implementation also allows for the train to transmit its location to the signaling system which could enable the use of moving or "virtual" blocks. The wireless implementation is generally cheaper in terms of equipment costs, but is considered to be much less reliable than using "harder" communications channels. For example, the wireless ITCS system on Amtrak's Michigan Line was still not functioning reliably in 2007 after 13 years of development, while the fixed ACSES system has been in daily service on the Northeast Corridor since 2002.

The fixed infrastructure method is proving popular on high-density passenger lines where pulse code cab signaling has already been installed. In some cases, the lack of a reliance on wireless communications is being touted as a benefit. The wireless method has proven most successful on low density, unsignaled dark territory normally controlled via track warrants, where speeds are already low and interruptions in the wireless connection to the train do not tend to compromise safety or train operations.

Some systems, like Amtrak's ACSES, operate with a hybrid technology that uses wireless links to update temporary speed restrictions or pass certain signals, with neither of these systems being critical for train operations.

## Locomotive Speed Control Unit

The equipment on board the locomotive must continually calculate the trains' current speed relative to a speed target some distance away governed by a braking curve. If the train risks not being able to slow to the speed target given the braking curve, the brakes are automatically applied and the train is immediately slowed. The speed targets are updated by information regarding fixed and dynamic speed limits determined by the track profile and signaling system.

Most current PTC implementations also use the speed control unit to store a database of track profiles attached to some sort of navigation system. The unit keeps track of the train's position along the rail line and automatically enforces any speed restrictions as well as the maximum authorized speed. Temporary speed restrictions can be updated before the train departs its terminal or via wireless data links. The track data can also be used to calculate braking curves based on the grade profile. The navigation system can use fixed track beacons or differential GPS stations combined with wheel rotation to accurately determine the train's location on the line within a few feet.

## Centralized Control

While some PTC systems interface directly with the existing signal system, others may maintain a set of vital computer systems at a central location that can keep track of trains and issue movement authorities to them directly via a wireless data network. This is often considered to be a form of Communications Based Train Control and is not a necessary part of PTC.

## Trackside Device Interface

The train may be able to detect the status of (and sometimes control) wayside devices, for example switch positions. This information is sent to the control center to further define the train's safe movements. Text messages and alarm conditions may also be automatically and manually exchanged between the train and the control center. Another capability would allow the employee-in-charge (EIC) to give trains permission to pass through their work zones via a wireless device instead of verbal communications.

## Technical Limitations

Even where safety systems such as cab signaling have been present for many decades, the freight railroad industry has been reluctant to fit speed control devices because the often heavy-handed nature of such devices can have an adverse effect on otherwise safe train operation. The advanced processor-based speed control algorithms found in PTC systems claim to be able to properly regulate the speed of freight trains over 5,000 feet (1,500 m) in length and weighing over 10,000 short tons (9,100 t), but concerns remain about taking the final decision out of the hands of skilled railroad engineers. Improper use of the air brake can lead to a train running away, derailment or to an unexpected separation.

Furthermore, an overly conservative PTC system runs the risk of slowing trains below the level at which they had previously been safely operated by human engineers. Railway speeds are calculated with a safety factor such that slight excesses in speed will not result in an accident. If a PTC system applies its own safety margin, then the end result will be an inefficient double safety factor. Moreover, a PTC system might be unable to account for variations in weather conditions or train handling, and might have to assume a worst-case scenario, further decreasing performance. In its 2009 regulatory filing, the FRA stated that PTC was in fact likely to decrease the capacity of freight railroads on many main lines. The European LOCOPROL/LOCOLOC project had shown that EGNOS-enhanced satellite navigation alone was unable to meet the SIL4 safety integrity required for train signaling.

From a purely technical standpoint, PTC will not prevent certain low-speed collisions caused by permissive block operation, accidents caused by trains "shoving" in reverse, derailments caused by track or train defect, grade crossing collisions, or collisions with previously derailed trains. Where

PTC is installed in the absence of track circuit blocks, it will not detect broken rails, flooded tracks, or dangerous debris fouling the line.

## Wireless Implementations

### Radio Spectrum Availability

The wireless infrastructure planned for use by all US Class I freights, most small freight railroads, and many commuter railroads is based on data radios operating in a single frequency band near 220 MHz. A consortium created by two freight railroads called PTC 220 LLC has purchased significant spectrum around 220 MHz, from previous licensees for use in deploying PTC. Some of this spectrum is in the form of nationwide licenses and some is not. The consortium plans to make this spectrum available for use by the US freights, but has indicated as recently as 2011 that they are unsure if they have enough spectrum to meet their needs. Several commuter railroads have begun purchasing 220 MHz spectrum in their geographic areas, but there is widespread concern that the acquisition of enough 220 MHz spectrum may be difficult to accomplish because of a lack of availability, difficulties in negotiating complex multi-party deals to gain enough adjacent spectrum, and because the financial cost of the acquisitions may make the task impossible for some state agencies. However, research suggests that dynamic spectrum allocation can solve the spectrum allocation problem at 220 MHz bandwidth.

Many of the railroads have requested that the FCC reallocate parts of the 220 MHz spectrum to them. They argue that they must have 220 MHz spectrum to be interoperable with each other. The FCC has stated that there is no reallocation forthcoming, that the railroads are not justified in requesting spectrum reallocation because they have not quantified how much spectrum they need, and that the railroads should seek spectrum in the secondary 220 MHz markets or in other bands.

### Radio Band

There are no regulatory or technical requirements that demand that 220 MHz be used to implement PTC (if a PTC implementation is to use wireless components at all). If wireless data transmission is necessary, there are a few advantages to the 220 MHz spectrum, provided it can be acquired at a reasonable cost. The first reason to consider using 220 MHz spectrum is PTC interoperability for freights and for some, but not all, commuter rail operations. Freight operations in the US often include the sharing of railroad tracks where one railroad's rail vehicles operate as a guest on another railroad's host tracks. Implementing PTC in such an environment is most easily achieved by using the same PTC equipment, and this includes radios and the associated radio spectrum.

When a commuter railroad operation must operate on a freight railroad territory, the commuter will most likely be required to install PTC equipment (including a radio) on their rail vehicle that is compliant with the freight railroad's PTC system, and this generally means the use of 220 MHz radios and spectrum. If the commuter uses the same PTC equipment, radios, and spectrum on their own property, they will be able to use it when their vehicles travel onto a freight's territory. From a practical standpoint, if the commuter instead elects to use another type of PTC on their own property, they will need to install a second set of onboard equipment so they can operate PTC on their own property while also operating PTC on a freight's property. If a multi-band radio (such as the current generation software defined radios) is not available, then separate radios and separate antennas will be necessary. With the complexity of track geometries, PTC requires a variable amount of the spectrum in a time critical manner. One way to achieve this is to extend the PTC software defined radios, such that it has the intelligence to allocate the spectrum dynamically. Adding the intelligence to the radio also helps to improve the security of the PTC communication medium.

If a small freight or commuter railroad does not operate on another railroad territory, then there is no interoperability-based reason that obligates them to use 220 MHz spectrum to implement PTC. In addition, if a small freight or commuter railroad only operates on their own territory and hosts other guest railroads (freight or other passenger rail), there is still no interoperability-based reason the host is obliged to use 220 MHz spectrum to implement PTC. Such a railroad could implement PTC by freely picking any radio spectrum and requiring the guest railroads to either install compliant PTC equipment (including radios) onboard their trains or provide wayside equipment for their guest PTC implementation to be installed on the host railroad property. An interesting case that highlights some of these issues is the northeast corridor. Amtrak operates services on two commuter rail properties it does not own: Metro-North Railroad (owned by New York and Connecticut) and Massachusetts Bay Transportation Authority (MBTA) (owned by Massachusetts). In theory, Amtrak could have found themselves installing their own PTC system on these host properties (about 15 percent of the corridor), or worse, found themselves in the ridiculous position of trying to install three different PTC systems on each Amtrak train to traverse the commuter properties. This was not the case. Amtrak had a significant head start over the commuter rail agencies on the corridor in implementing PTC. They spent a considerable amount of time in research and development and won early approvals for their ACSES system on the northeast corridor with the FRA. They chose first to use 900 MHz and then later moved to 220 MHz, in part because of a perceived improvement in radio-system performance and in part because Amtrak was using 220 MHz in Michigan for their ITCS implementation.

When the commuter agencies on the corridor looked at options for implementing PTC, many of them chose to take advantage of the advance work Amtrak had done and implement the ACSES solution using 220 MHz. Amtrak's early work paid off and meant that they would be traversing commuter properties that installed the same protocol at the same frequency, making them all interoperable. (Actually most of the Northeast Corridor is owned and operated by Amtrak, not the commuter properties, including the tracks from Washington, D.C. to New York Penn Station and the tracks from Philadelphia to Harrisburg, Pennsylvania. The State of Massachusetts owns the tracks from the Rhode Island state line to the New Hampshire state line, but Amtrak "operates" these lines. Only the line between New York City and New Haven, Connecticut is actually owned and operated by a commuter line.)

One other perceived reason to consider 220 MHz for PTC may be PTC-compatible radio equipment availability. Radio equipment specifically targeted toward PTC is currently only available from a limited number of vendors, and they are focused only on 220 MHz. One radio vendor in particular, Meteorcomm LLC, is able to support the I-ETMS PTC protocol with a 220 MHz radio. Meteorcomm is jointly owned by several of the Class I freights, and some in the industry have indicated that using their 220 MHz radio and associated equipment will be done on a per-site licensing basis. Recurring fees may be associated with this process too. There is further concern that the 'buy in' and licensing fees will be significant, and this has led some to speculate that the owners of Meteorcomm (the freights) may have legal exposure to anti-trust violations. For many railroads, there is no other practical option to meet the federal mandate than to install PTC at 220 MHz using I-ETMS with the Meteorcomm radios. On the northeast corridor, another radio vendor, GE MDS, is able to support the Amtrak ACSES protocol with a 220 MHz radio. It should be stressed that the main concern among the freights regarding the PTC deadline is the availability of PTC equipment. With an eye to anti-trust issues and ready radio availability, Meteorcomm radio designs have been second-sourced to CalAmp radios. This all may mean that there is not enough 220 MHz PTC radio equipment available for all of the railroads that must implement PTC.

There are also issues with the use of these frequencies outside the US; in Canada, 220 MHz remains part of the radioamateur 1.25-metre band.

Other bands besides 220 MHz will support PTC, and have been used to win approvals from the FRA for PTC. When Amtrak received their initial approval, they planned to use 900 MHz frequencies for ACSES. BNSF Railway won its first PTC approvals from the FRA for an early version of ETMS using a multi-band radio that included 45 MHz frequencies, 160 MHz frequencies, 900 MHz frequencies and WiFi. A small freight or commuter that selects one or more of these bands or another one such as 450 MHz might find it easier to acquire spectrum. They will need to research spectrum issues, radio equipment, antennas, and protocol compatibility issues to successfully deploy PTC.

## Interoperability Requirements

There is no single defined standard for “interoperable PTC systems”. Several examples of interoperable systems illustrate this point. First, the UP and BNSF are interoperable across their systems. They are both implementing I-ETMS and will use different radio frequencies in different locations. In the second example, Amtrak is interoperable with Norfolk Southern in Michigan. Amtrak uses ITCS, while Norfolk Southern uses I-ETMS. To interoperate, two 220 MHz radios are installed in each wayside location and they both interface with a common PTC system through an interface device (similar to a network gateway or protocol converter) at each wayside location. One radio talks to freight trains using I-ETMS and one radio talks to passenger trains using ITCS. In this case interoperability stops at the wayside and does not include the wireless segment out to the rail vehicles or the onboard systems. In the third example, similar to the first, Metrolink, the commuter rail agency in Los Angeles, is implementing I-ETMS and will use the same PTC equipment as both the UP and BNSF. Metrolink is procuring their own 220 MHz spectrum so that trains on Metrolink territory (commuter and freight) will use other channels than those used by the UP and BNSF. Interoperability is achieved by directing the onboard radio to change channels depending on location. For SEPTA, the commuter operation in and around Philadelphia, Ansaldo is implementing ACSES, the Amtrak northeast corridor PTC protocol. For CSX all the ACSES PTC transactions will be handed to CSX at the SEPTA back office, and CSX will be responsible for deploying I-ETMS infrastructure that they will use to communicate with their freight trains. The SEPTA interoperability model is very similar to that of the public safety radio community wherein different radio systems that use different frequencies and protocols are cross-connected only in the back office to support system to system communications.

## Multi-band Solutions

For the major freight railroads and Amtrak the answer seems to be that one frequency band is sufficient. These rail operations measure on-time performance on a much more coarse scale than commuters do so their tolerance for delay is greater and has less impact on train schedules. In addition, the PTC implementations deployed by commuter operations will be running much closer to the performance envelope than that of either Amtrak or the freights. For commuters in particular there is therefore some concern that implementing PTC with a single frequency band may not be sufficient. The single frequency-band approach to supporting real-time train control has a history of being difficult to use for such applications. This difficulty is not unique to train control. Interference, both man-made and natural, can at times affect the operation of any wireless system that relies on one frequency band. When such wireless systems are employed for real-time control networks it is very difficult to ensure that network performance will not sometimes be impacted. CSX encountered this problem when it experienced propagation ducting problems in its 900 MHz Advanced Train Control System (ATCS) network in the 1990s. The ATCS protocol, which the AAR had recommended the FCC consider as PTC in 2000 (when AAR sought a nationwide 900 MHz "ribbon" license), can support train control operation at both 900 MHz and 160 MHz. The latter frequency band is only used for ATCS on a few subdivisions and shortlines. More recently, the industry had been moving toward a more robust multi-band radio solution for data applications such as PTC. In 2007, BNSF first won FRA approval for their original ETMS PTC system using a multi frequency-band radio. In addition, in mid-2008, an FRA sponsored effort by the AAR to develop a Higher Performance Data Radio (HPDR) for use at 160 MHz actually resulted in a contract being awarded to Meteorcomm for a 4-band radio to be used for voice and data. These more recent multi-band radio efforts were shelved in late 2008, after the Rail Safety Improvements Act became law, and the freights decided to pursue PTC using 220 MHz alone, in a single frequency-band configuration. Amtrak and most commuter operations quickly followed suit, selecting 220 MHz.

## Suitability of Wireless PTC for Commuter Rail

Soon after the Rail Safety Improvements Act was passed, many commuter railroads chose not to develop their own PTC protocol and instead decided to save time and money by using a protocol developed for either freight or long haul passenger (Amtrak) operations. Deploying such a protocol for urban commuter operation, where it will be necessary to support numerous, small, fast-moving trains, will be a challenge. It remains to be seen whether the performance envelope of PTC protocols developed and optimized for less numerous, slower and/or larger trains can support a more complex operational scenario, such as that of a commuter rail operation, without impacting on-time performance. Detailed and exhaustive protocol simulation testing can ease the risk of problems, however, there are too many variables, especially when the wireless component is considered, to guarantee before hand that under certain worst-case operational profiles in certain locations, train operations will not be impacted. In fact, during system acceptance testing, such worst-case operational profiles may not even be tested because of the effort involved. One need only consider what it would take to identify the PTC protocol train capacity limitations at each interlocking of a large commuter rail operation when a train is broken down at the interlocking and 10–20 other trains are within communications range of a single wayside location. Such a what-if scenario may be tested at a few interlockings but not at the 30 or more interlockings on a large commuter property.

## Open Standards

A large group of industry experts from the federal government, manufacturers, railroads, and consultants are participating in a study group sponsored by the IEEE 802.15 working group, to look at using lessons learned in protocol development in the IEEE 802 suite to propose a comprehensive solution to the wireless component of PTC. While this effort may not significantly change the current United States PTC efforts already underway, an open standard could possibly provide a way forward for all of the railroads to eventually deploy a more interoperable, robust, reliable, future-proof, and scalable solution for the wireless component of PTC.

## Upgrade Costs

The railroad industry, like the process industry and the power utility industry, has always demanded that the return on investment for large capital investments associated with infrastructure improvements be fully realized before the asset is decommissioned and replaced. This paradigm will be applied to PTC as well. It is highly unlikely that there will be any major upgrades to initial PTC deployments within even the first 10 years. The calculation for return on investment is not a simple one and some railroads may determine, for instance after five years, that an upgrade of certain components of PTC may be justified. An example could be the radio component of PTC. If an open standard creates a less expensive radio product that is backwards compatible to existing systems and that perhaps improves PTC system performance and also includes improvements that save on operational costs, then a railroad would be prudent to consider a plan for replacing their PTC radios.

## References


- Kakodkar, Anil; Sreedharan, E.; Vedachalam, N.; Dhande, Sanjay G.; Srivastava, G.P.; Amitabh (February 2012). Report of High Level Safety Review Committee (Report). New Delhi: Government of India Ministry of Railways. Retrieved 11 October 2014
- Train-protection, signaling: railway-technical.com, Retrieved 12 April, 2019
- Bandara, Damindra; Abadie, Andre; Wijesekara, Duminda (2015). Cell planning for high-speed train operations in USA. Proceedings of the 2015 Joint Rail Conference. Doi:10.1115/JRC2015-5805
- Automatic-train-control, signaling: railway-technical.com, Retrieved 13 May, 2019
- "Positive Train Control | Federal Railroad Administration". Www.fra.dot.gov. Department of Transportation. Retrieved 12 February 2018

# Permissions

All chapters in this book are published with permission under the Creative Commons Attribution Share Alike License or equivalent. Every chapter published in this book has been scrutinized by our experts. Their significance has been extensively debated. The topics covered herein carry significant information for a comprehensive understanding. They may even be implemented as practical applications or may be referred to as a beginning point for further studies.

We would like to thank the editorial team for lending their expertise to make the book truly unique. They have played a crucial role in the development of this book. Without their invaluable contributions this book wouldn't have been possible. They have made vital efforts to compile up to date information on the varied aspects of this subject to make this book a valuable addition to the collection of many professionals and students.

This book was conceptualized with the vision of imparting up-to-date and integrated information in this field. To ensure the same, a matchless editorial board was set up. Every individual on the board went through rigorous rounds of assessment to prove their worth. After which they invested a large part of their time researching and compiling the most relevant data for our readers.

The editorial board has been involved in producing this book since its inception. They have spent rigorous hours researching and exploring the diverse topics which have resulted in the successful publishing of this book. They have passed on their knowledge of decades through this book. To expedite this challenging task, the publisher supported the team at every step. A small team of assistant editors was also appointed to further simplify the editing procedure and attain best results for the readers.

Apart from the editorial board, the designing team has also invested a significant amount of their time in understanding the subject and creating the most relevant covers. They scrutinized every image to scout for the most suitable representation of the subject and create an appropriate cover for the book.

The publishing team has been an ardent support to the editorial, designing and production team. Their endless efforts to recruit the best for this project, has resulted in the accomplishment of this book. They are a veteran in the field of academics and their pool of knowledge is as vast as their experience in printing. Their expertise and guidance has proved useful at every step. Their uncompromising quality standards have made this book an exceptional effort. Their encouragement from time to time has been an inspiration for everyone.

The publisher and the editorial board hope that this book will prove to be a valuable piece of knowledge for students, practitioners and scholars across the globe.

# Index

www.ingramcontent.com/pod-product-compliance
Lightning Source LLC
LaVergne TN
LVHW080707210926
848485LV00003B/428